Lectures on QED and QCD

Practical Calculation and Renormalization of
One- and Multi-Loop Feynman Diagrams

Lectures on QED and QCD

Practical Calculation and Renormalization of
One- and Multi-Loop Feynman Diagrams

Andrey Grozin
Budker Institute of Nuclear Physics, Russia

World Scientific

NEW JERSEY · LONDON · SINGAPORE · BEIJING · SHANGHAI · HONG KONG · TAIPEI · CHENNAI

Published by

World Scientific Publishing Co. Pte. Ltd.

5 Toh Tuck Link, Singapore 596224

USA office: 27 Warren Street, Suite 401-402, Hackensack, NJ 07601

UK office: 57 Shelton Street, Covent Garden, London WC2H 9HE

British Library Cataloguing-in-Publication Data
A catalogue record for this book is available from the British Library.

LECTURES ON QED AND QCD

Copyright © 2007 by World Scientific Publishing Co. Pte. Ltd.

All rights reserved. This book, or parts thereof, may not be reproduced in any form or by any means, electronic or mechanical, including photocopying, recording or any information storage and retrieval system now known or to be invented, without written permission from the Publisher.

For photocopying of material in this volume, please pay a copying fee through the Copyright Clearance Center, Inc., 222 Rosewood Drive, Danvers, MA 01923, USA. In this case permission to photocopy is not required from the publisher.

ISBN 978-981-256-914-1

Printed in Singapore

Preface

Precision of experimental data in many areas of elementary particle physics is quickly improving. For example, the anomalous magnetic moment of the muon is measured with a fantastically high precision. Results from B factories at SLAC and KEK for many quantities have low systematic errors and very high statistics. Of course, there are many more examples of such progress.

To compare high-precision experimental data with the theory, one has to obtain equally high-precision theoretical expressions for the measured quantities. Preparing physical programs for future colliders also requires high-precision theoretical calculations. In order to be able to search for a new physics, one has to understand standard processes (which can be a source of background) at a highly detailed level.

This means, in particular, calculation of higher radiative corrections. They are described by Feynman diagrams with one or several loops. Calculation of such diagrams is a very non trivial task. It involves solving deep mathematical problems. Even when a suitable calculation algorithm has been constructed, this is not the end of the story. Often, many thousands of diagrams have to be calculated. This requires an unprecedented level of automation of theoretical research: generation and calculation of the diagrams have to be done systematically, by computer programs, without any interference of a human researcher. Some of the calculations of this kind are among the largest computer-algebraic calculations ever performed. This area of theoretical physics is progressing rapidly. A large number of physicists in many countries are involved in such activities. And this number is increasing. Many of today's students in the area of theoretical high energy physics will be involved in calculations of radiative corrections in the course of their careers.

Quantum field theory textbooks usually don't describe methods of calculation of multiloop Feynman diagrams. Most textbooks discuss quantization of fields (including gauge theories), obtain Feynman rules, and show a few simple examples of one-loop calculations. On the other hand, there is a huge amount of literature for experts in the area of multiloop calculations, usually in the form of original papers and specialized review articles. The purpose of this book is to close the gap between textbooks and the modern research literature. The reader should have a firm grasp of the basics of quantum field theory, including quantization of gauge fields (Faddeev–Popov ghosts, etc.) and Feynman rules. These topics can be found in any modern textbook, e.g., in [Peskin and Schroeder (1995)]. No previous experience in calculating Feynman diagrams with loops is required. Fundamental concepts and methods used for such calculations, as well as a large number of examples, are presented in this book in detail.

The main focus of the book is on quantum electrodynamics (QED) and quantum chromodynamics (QCD). In the area of QED, some extremely high-precision experimental data are available (anomalous magnetic moments, hydrogen atom, positronium). Correspondingly, some groundbreaking theoretical calculations have been done. In the area of QCD, very high precision comparisons of the theory and experiments are never possible, because we still don't know how to take non-perturbative phenomena into account quantitatively and in a model-free way (except by lattice Monte–Carlo simulations, whose accuracy is not very high but is increasing). However, the QCD coupling constant is much larger, and several terms of perturbative series are usually required to obtain the necessary (moderate) precision. Calculations in QED and QCD are usually very similar, but QCD ones are more lengthy — more diagrams, colour factors, etc. Therefore, a large fraction of the text is (technically) devoted to QED, but it should be considered also as a demonstration of methods which are used in QCD.

The first part of this book is based on lectures given to students preparing for the M. Sc. degree at Dubna International Advanced School on Theoretical Physics in 2005 and at Universität Karlsruhe. They were published as hep-ph/0508242. They were revised and extended for this book.

Practically all modern multiloop calculations are performed in the framework of dimensional regularization. It is discussed in Chap. 1, together with simple (but fundamentally important) one-loop examples. In Chaps. 2 and 3, one-loop corrections in QED and QCD are discussed. Here we use the $\overline{\text{MS}}$ renormalization scheme, which is most popular, especially in

QCD. Methods and results of calculation of two-loop corrections in QED and QCD are introduced in Chap. 4, also using $\overline{\text{MS}}$ scheme. Chap. 5 is devoted to the on-shell renormalization scheme, which is most often used in QED at low energies, but also for heavy-quark problems in QCD. Decoupling of heavy quarks is most fundamental in QCD; it is employed practically every time one does any work in QCD. It is presented in Chap. 6, where a simplified QED problem is considered in detail; it makes understanding the problem much easier. This is the first time decoupling in the $\overline{\text{MS}}$ scheme is considered in a textbook, with full calculations presented. Finally, Appendix A is a practical guide on calculating colour factors, which is a necessary (though simple) step in any QCD work. Here I follow an excellent book [Cvitanović (web-book)] available on the Web.

The second part is based on lectures given to Ph. D. students at the International School "Calculations at modern and future colliders", Dubna (2003), and at Universität Karlsruhe. They were published in Int. J. Mod. Phys. A **19** (2004) 473. They are (slightly) revised for this book. This second lecture course forms a natural sequel to the main one. It discusses some advanced methods of multiloop calculations; in cases when the same problem is discussed in both courses, it is solved by different methods. So, studying both of the courses gives a wider perspective and a better toolbox of methods. For a much more comprehensive presentation of modern methods of calculating Feynman integrals, the reader is addressed to a recent book [Smirnov (2006)].

Of course, there are a lot of things which are *not* discussed in this book. It only shows the most simple and fundamental examples. More complicated scattering processes (diagrams with more external legs) and radiative corrections in the electroweak theory (which often involve several kinds of particles with different masses) are not considered here. But the general approaches (dimensional regularization, $\overline{\text{MS}}$ renormalization, integration by parts...) remain the same. After reading this book, the reader should have no problems reading specialized literature about more advanced problems.

I am grateful to D.J. Broadhurst, K.G. Chetyrkin, A. Czarnecki, A.I. Davydychev, A.V. Smirnov, V.A. Smirnov for collaboration on various multiloop projects and numerous discussions, and to the organizers of the Dubna schools D.I. Kazakov, S.V. Mikhailov, A.A. Vladimirov for inviting me to give the lectures and for advices on the contents. A large part of the work on the book was done at the University of Karlsruhe, and was supported by DFG through SFB/TR 9; I am grateful to J.H. Kühn and M. Steinhauser for inviting me to Karlsruhe and fruitful discussions.

Andrey Grozin

Contents

PART 1
QED and QCD

Chapter 1

One-loop diagrams

1.1 Divergences, regularization and renormalization

When interactions in a quantum field theory may be considered weak, we can use perturbation theory, starting from the theory of free fields in zeroth approximation. Contributions to perturbative series can be conveniently depicted as Feynman diagrams; corresponding analytical expressions can be reconstructed from the diagrams using Feynman rules. If a Feynman diagram contains a loop (or several loops), its expression contains an integral over the loop momentum (or several loop momenta). Such integrals often diverge at large loop momenta (*ultraviolet divergences*). For example, let's consider the scalar field theory with the $g\varphi^3$ interaction. The one-loop correction to the propagator (Fig. 1.1) is

$$g^2 \int \frac{d^4k}{[m^2 - k^2 - i0]\,[m^2 - (p+k)^2 - i0]}\,.\qquad(1.1)$$

At $k \to \infty$, the denominator behaves as k^4, and the integral diverges logarithmically.

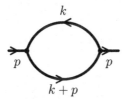

Fig. 1.1   One-loop propagator diagram

Therefore, first of all, we need to introduce a regularization — some modification of the theory which makes loop integrals convergent. One can sensibly manipulate and calculate regularized Feynman integrals. In the physical limit the original theory is restored. Then we should re-formulate the problem. The original perturbative expression (which contains divergent Feynman integrals, and hence is senseless) expresses a scattering amplitude via bare masses and charges which are present in the Lagrangian. But physical masses and charges don't coincide with these bare quantities, if we take radiative corrections into account. Therefore, we should fix some definition of physical masses and charges, and re-express our scattering amplitude via these quantities. This procedure is called renormalization. It is physically necessary, independently of the problem of divergences. If the theory makes sense, expressions for scattering amplitudes via physical masses and charges will remain finite when regularization is removed.

Of course, many different regularization methods can be invented. For example, a cutoff can be introduced into loop integrals by replacing propagators:

$$\frac{1}{m^2 - k^2 - i0} \to \frac{\theta\left(|k^2| < \Lambda^2\right)}{m^2 - k^2 - i0}.$$

The physical limit is $\Lambda \to \infty$. However, such a cutoff makes calculation of diagrams extremely difficult, because the integration region has a complicated shape. In addition to this, integrating by parts becomes very complicated because of boundary terms. Pauli–Villars regularization

$$\frac{1}{m^2 - k^2 - i0} \to \frac{1}{m^2 - k^2 - i0} - \frac{1}{M^2 - k^2 - i0}$$

(with the physical limit $M \to \infty$) is much better. However, it is not very good for gauge theories: gauge bosons in an (unbroken) gauge theory must be massless, and modifying their propagators by introducing massive terms breaks the gauge invariance.

In general, a good regularization method should preserve simple rules for manipulating loop integrals (like integration by parts), and also should preserve as much of symmetries of the theory as possible. Unbroken symmetries make calculations much simpler by restricting possible form of results. Sometimes, it is not possible to preserve *all* symmetries of a field theory when performing its regularization. In such a case, it may happen that renormalized results break some symmetry even in the limit of no regularization. This means that the quantum field theory has less symmetries than its classical Lagrangian suggests (an *anomaly*).

A popular regularization of gauge theories (after analytic continuation to Euclidean space-time) is to replace the continuous space-time by a cubic lattice with spacing a. The physical limit of this regularization is $a \to 0$. It can be done in an exactly gauge-invariant way invented by Wilson (matter fields live at lattice points, and gauge fields live on one-dimensional links). Within this approach, quantitative results can be obtained by Monte–Carlo simulation, without relying on perturbation theory. However, this regularization breaks Lorentz invariance (only a smaller symmetry group, that of a 4-dimensional cube, is preserved). This makes perturbative calculations much more difficult.

The most popular method used in multiloop calculations nowadays is *dimensional regularization*. Diagrams are calculated in d-dimensional space-time. The dimensionality d must appear in all formulas as a symbol, it is not enough to obtain separate results for a few integer values of d. The physical limit is $d \to 4$; therefore, d is often written as $4 - 2\varepsilon$. Divergences in intermediate perturbative formulas appear as $1/\varepsilon$ poles. After calculating a physical result in terms of physical parameters, we can take the limit $\varepsilon \to 0$ (this limit should exist in a sensible theory).

Dimensional regularization allows one to use simple algebraic rules for manipulating Feynman integrals. In particular, all integrations are over the whole infinite momentum space, and no surface terms appear during integration by parts. Dimensional regularization preserves Lorentz invariance (making it d-dimensional; when we take the limit $\varepsilon \to 0$ at the end of calculations, results automatically have a 4-dimensionally Lorentz-invariant form). In gauge theories, the d-dimensional Lagrangian is gauge invariant, so, the symmetry is preserved. Most other symmetries are preserved, too.

However, there are exceptions. As we shall see in Sect. 1.6, the Dirac matrix γ_5 cannot be generalized to d dimensions. Therefore, if we have a theory with massless fermions having chiral symmetry, this symmetry is not preserved in d dimensions, and in some cases it may be broken in final renormalized results at $\varepsilon \to 0$ (*axial anomaly*). Also, continuation to d dimensions changes dimensionalities of various quantities. Therefore, if the 4-dimensional massless theory was scale invariant, this symmetry (and a more general conformal symmetry) will be broken by regularization. This breaking can persist in renormalized results at $\varepsilon \to 0$ (*conformal anomaly*). Another important symmetry which is broken by dimensional regularization is supersymmetry. In supersymmetric theories, the numbers of bosonic and fermionic degrees of freedom coincide. However, these numbers depend on d in different ways, and supersymmetry is broken at $d \neq 4$.

Until now, we discussed ultraviolet divergences. In theories with massless particles (for example, gauge theories) some diagrams can also diverge at $k \to 0$ (*infrared divergences*)[1]. They cannot appear in results for meaningful physical quantities (we cannot detect arbitrarily soft photons, so, cross sections should be summed over final states with any number of such photons). In order to do intermediate manipulations, we have to regularize infrared divergences, too. This can be done by introducing a small photon mass; however, such a regularization breaks gauge invariance. Dimensional regularization regularizes infrared divergences as well as ultraviolet ones: both appear as $1/\varepsilon$ poles (in general, it is very difficult to trace which $1/\varepsilon$ poles are of ultraviolet origin and which are infrared).

1.2   Massive vacuum diagram

So, during these lectures, we are going to live in d-dimensional space–time: one time and $d - 1$ space dimensions.

Fig. 1.2   One-loop massive vacuum diagram

Let's consider the simplest diagram shown in Fig. 1.2:

$$\int \frac{d^d k}{D^n} = i\pi^{d/2} m^{d-2n} V(n) , \quad D = m^2 - k^2 - i0 . \tag{1.2}$$

The power of m is evident from the dimensional counting, and our aim is to find the dimensionless function $V(n)$; we can put $m = 1$ to simplify the calculation. The poles in the complex k_0 plane are situated at

$$k_0 = \pm \left(\sqrt{\vec{k}^2 + 1} - i0 \right) \tag{1.3}$$

---

[1]If there is an on-shell massless particle with momentum p in the process, there can also be *collinear divergences* when the momentum k of a virtual particle is non-zero but parallel to p.

(Fig. 1.3). The integration contour C is initially along the real axis. We may rotate it counterclockwise by $\pi/2$ without crossing the poles. After this Wick rotation, we integrate along the imaginary axis in the k_0 plane in the positive direction: $k_0 = i\boldsymbol{k}_0$. Here $\boldsymbol{k}_0$ is the 0-th component of the vector $\boldsymbol{k}$ in d-dimensional Euclidean space (Euclidean vectors will be denoted by the bold font), and $k^2 = -\boldsymbol{k}^2$. Then our definition (1.2) of $V(n)$ becomes

$$\int \frac{d^d\boldsymbol{k}}{(\boldsymbol{k}^2 + 1)^n} = \pi^{d/2}V(n). \tag{1.4}$$

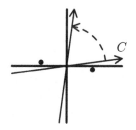

Fig. 1.3   Complex k_0 plane

It is often useful to turn denominators into exponentials using the α-parametrization

$$\frac{1}{a^n} = \frac{1}{\Gamma(n)} \int_0^\infty e^{-a\alpha}\alpha^{n-1}d\alpha. \tag{1.5}$$

For our integral, this gives

$$V(n) = \frac{\pi^{-d/2}}{\Gamma(n)} \int e^{-\alpha(\boldsymbol{k}^2+1)}\alpha^{n-1}d\alpha\, d^d\boldsymbol{k}. \tag{1.6}$$

The d-dimensional integral of the exponent of the quadratic form is the product of d one-dimensional integrals:

$$\int e^{-\alpha \boldsymbol{k}^2}d^d\boldsymbol{k} = \left[\int_{-\infty}^{+\infty} e^{-\alpha \boldsymbol{k}_x^2}dk_x\right]^d = \left(\frac{\pi}{\alpha}\right)^{d/2}. \tag{1.7}$$

This is the definition of d-dimensional integration; note that the result contains d as a symbol. Now it is easy to calculate

$$V(n) = \frac{1}{\Gamma(n)} \int_0^\infty e^{-\alpha}\alpha^{n-d/2-1}d\alpha. \tag{1.8}$$

The denominator in (1.4) behaves as $(\boldsymbol{k}^2)^n$ at $\boldsymbol{k} \to \infty$. Therefore, the integral converges only if $d < 2n$ (or $\operatorname{Re} d < 2n$ if we also consider complex d). This is also clear from (1.8). In this region, it can be calculated via Γ-function:

$$V(n) = \frac{\Gamma(-d/2 + n)}{\Gamma(n)}. \tag{1.9}$$

We define the integral in the whole complex plane of d as the analytic continuation from the region where it converges. In this particular case, this means the formula (1.9) everywhere.

For all integer n, the results are proportional to

$$V_1 = V(1) = \frac{4}{(d-2)(d-4)}\Gamma(1+\varepsilon), \tag{1.10}$$

where the coefficients are rational functions of d. For example,

$$V(2) = -\frac{d-2}{2}V_1 = \Gamma(\varepsilon). \tag{1.11}$$

The integral $V(n)$ is ultraviolet divergent at $d \to 4$ if $n \leq 2$. This ultraviolet divergence shows itself as a $1/\varepsilon$ pole in (1.9) for $n = 1, 2$.

Results for Feynman integrals in d dimensions often contain Γ-functions of arguments $n + m\varepsilon$. In order to expand such results in ε, we first reduce then to $\Gamma(1 + m\varepsilon)$, and then use

$$\Gamma(1+\varepsilon) = \exp\left[-\gamma\varepsilon + \sum_{n=2}^{\infty} \frac{(-1)^n \zeta_n}{n}\varepsilon^n\right], \tag{1.12}$$

where γ is the Euler constant, and

$$\zeta_n = \sum_{k=1}^{\infty} \frac{1}{k^n} \tag{1.13}$$

is the Riemann ζ-function:

$$\zeta_2 = \frac{\pi^2}{6}, \quad \zeta_3 \approx 1.202, \quad \zeta_4 = \frac{\pi^4}{90}, \quad \cdots \tag{1.14}$$

1.3 Integrals in d dimensions

Integrals in d dimensions have the usual properties:

$$\int cf(k)\, d^d k = c \int f(k)\, d^d k \,, \tag{1.15}$$

$$\int [f(k) + g(k)]\, d^d k = \int f(k)\, d^d k + \int g(k)\, d^d k \,, \tag{1.16}$$

$$\int f(k+q)\, d^d k = \int f(k)\, d^d k \,, \tag{1.17}$$

$$\int f(\Lambda k)\, d^d k = \int f(k)\, d^d k \,, \tag{1.18}$$

$$\int f(ck)\, d^d k = c^{-d} \int f(k)\, d^d k \,, \tag{1.19}$$

where $k'^\mu = \Lambda^\mu{}_\nu k^\nu$ is a Lorentz transformation. In particular, the translational invariance (1.17) for an infinitesimal q gives a very important property

$$\int \frac{\partial f(k)}{\partial k^\mu}\, d^d k = 0 \,, \tag{1.20}$$

which we shall often use. It can be understood directly: in the infinite d-dimensional momentum space there can be no boundary terms. Similarly, an infinitesimal Lorentz transformation is $\Lambda_{\mu\nu} = g_{\mu\nu} + \varepsilon_{\mu\nu}$, where $\varepsilon_{\mu\nu} = -\varepsilon_{\nu\mu}$ is an infinitesimal antisymmetric tensor. Therefore, from (1.18),

$$\int \left(k_\mu \frac{\partial}{\partial k^\nu} - k_\nu \frac{\partial}{\partial k^\mu} \right) f(k)\, d^d k = 0 \,. \tag{1.21}$$

This Lorentz-invariance condition also can be used for deriving useful relations among Feynman integrals.

What is the value of the massless vacuum diagram

$$\int \frac{d^d k}{(-k^2 - i0)^n}$$

(Fig. 1.4)? Its dimensionality is $d - 2n$. But it contains no dimensionful parameters from which such a value could be constructed. The only result we can write for this diagram is

$$\int \frac{d^d k}{(-k^2 - i0)^n} = 0 \,. \tag{1.22}$$

Fig. 1.4   One-loop massless vacuum diagram

Equivalently, from (1.19),

$$\int \frac{d^d k}{(-k^2 - i0)^n} = \int \frac{d^d c k}{[-(ck)^2 - i0]^n} = c^{d-2n} \int \frac{d^d k}{(-k^2 - i0)^n} ,$$

and we obtain (1.22). This argument fails at $n = d/2$; surprises can be expected at this point.

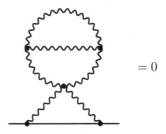

Fig. 1.5   A diagram for the quark propagator

This dimensions-counting argument is much more general. For example, the diagram in Fig. 1.5 contains a sub-diagram which is attached to the rest of the diagram at a single vertex, and which contains no scale. This subdiagram is given by an integral (maybe, a tensor one) with some d-dependent dimensionality. It has no dimensionful parameters; the only value we can construct for such an integral is 0.

A typical l-loop Feynman integral contains n denominators of propagators (which are quadratic in momenta) and, maybe, a numerators (polynomial in momenta). Using α-parametrization (1.5), we can write it as an integral over n parameters α_i of an integral over l Euclidean d-dimensional loop momenta $\boldsymbol{k}_i$ (after Wick rotation) of an exponent of a quadratic form (maybe, multiplied by the polynomial numerator). Loop momenta in the numerator can be replaced by derivatives acting on the exponent. Terms in the exponent linear in $\boldsymbol{k}_i$ can be eliminated by shifting integration momenta. We are left with the integral

$$\int e^{-\sum A_{ij}\boldsymbol{k}_i\cdot\boldsymbol{k}_j} \prod d^d\boldsymbol{k}_i \,,$$

where A_{ij} is a symmetric matrix. But this is the product of d copies of integrals involving separate components of the integration momenta:

$$\int e^{-\sum A_{ij}\boldsymbol{k}_i\cdot\boldsymbol{k}_j} \prod d^d\boldsymbol{k}_i = \left(\int e^{-\sum A_{ij}k_{ix}k_{jx}} \prod dk_{ix}\right)^d .$$

This integral is easy to calculate:

$$\int e^{-\sum A_{ij}k_i k_j} \prod dk_i = \left(\frac{\pi^l}{\det A}\right)^{1/2} .$$

This formula becomes obvious in the basis of eigenvectors of A, where A is diagonal and $\det A$ is the product of its eigenvalues. We arrive at

$$\int e^{-\sum A_{ij}\boldsymbol{k}_i\cdot\boldsymbol{k}_j} \prod d^d\boldsymbol{k}_i = \left(\frac{\pi^l}{\det A}\right)^{d/2} . \tag{1.23}$$

This is a constructive definition of d-dimensional integration which generalizes (1.7).

With this definition, a seemingly obvious property

$$\int \frac{f(\boldsymbol{k}_i)g(\boldsymbol{k}_i)}{g(\boldsymbol{k}_i)} \prod d^d\boldsymbol{k}_i = \int f(\boldsymbol{k}_i) \prod d^d\boldsymbol{k}_i \tag{1.24}$$

(where $g(\boldsymbol{k}_i)$ is quadratic in momenta) becomes a theorem which needs a proof. Let's consider

$$\int \frac{\boldsymbol{k}^2 + m^2}{\boldsymbol{k}^2 + m^2} f(\boldsymbol{k}) \, d^d\boldsymbol{k}$$

(the momentum present in the denominator g is called $\boldsymbol{k}$; integrations in other loop momenta, if present, are irrelevant). This integral is (see (1.5))

$$\int d^d\boldsymbol{k}\, f(\boldsymbol{k}) \int_0^\infty d\alpha\, (\boldsymbol{k}^2 + m^2) e^{-\alpha(\boldsymbol{k}^2+m^2)}$$

$$= -\int d^d\boldsymbol{k}\, f(\boldsymbol{k}) \int_0^\infty d\alpha\, \frac{\partial}{\partial\alpha} e^{-\alpha(\boldsymbol{k}^2+m^2)} = \int d^d\boldsymbol{k}\, f(\boldsymbol{k}) \,.$$

Let's calculate the full solid angle in d-dimensional space. To this end, we'll calculate one and the same integral in Cartesian coordinates

$$\int e^{-\boldsymbol{k}^2} d^d\boldsymbol{k} = \left[\int_{-\infty}^{+\infty} e^{-k_x^2} dk_x \right]^d = \pi^{d/2} \qquad (1.25)$$

and in spherical coordinates

$$\Omega_d \int_0^\infty e^{-\boldsymbol{k}^2} k^{d-1} dk = \frac{\Omega_d}{2} \int_0^\infty e^{-\boldsymbol{k}^2} (\boldsymbol{k}^2)^{d/2-1} d\boldsymbol{k}^2 = \frac{\Omega_d \Gamma(d/2)}{2}. \qquad (1.26)$$

Therefore, the full solid angle is

$$\Omega_d = \frac{2\pi^{d/2}}{\Gamma(d/2)}. \qquad (1.27)$$

For example,

$$\Omega_1 = 2, \quad \Omega_2 = 2\pi, \quad \Omega_3 = 4\pi, \quad \Omega_4 = 2\pi^2, \quad \ldots \qquad (1.28)$$

In one-dimensional space, a sphere consists of 2 points, hence $\Omega_1 = 2$; the values for $d = 2$ and 3 are also well known.

The integral

$$\int \frac{d^d k}{(2\pi)^d} \frac{1}{(-k^2 - i0)^2}$$

contains both an ultraviolet and an infrared divergences; they cancel each other, and the integral vanishes. Suppose we are interested only in the ultraviolet divergence $(1/\varepsilon)$ of this integral. It depends only on the behaviour of the integrand at $k \to \infty$. We may choose any infrared regularization in order to remove the infrared divergence of this integral. The simplest possibility is to introduce a sharp infrared cutoff into the Euclidean integral (we may use Ω_4 (1.28) here)

$$\int \frac{d^d k}{(2\pi)^d} \frac{1}{(-k^2)^2} \bigg|_{UV} = \frac{i}{8\pi^2} \int_\lambda^\infty k^{-1-2\varepsilon} dk = \frac{i\lambda^{-2\varepsilon}}{(4\pi)^2 \varepsilon} = \frac{i}{(4\pi)^2} \frac{1}{\varepsilon}. \qquad (1.29)$$

Any infrared regularization can be used; instead of a cutoff, we could insert a non-zero mass, for example:

$$\int \frac{d^d k}{(2\pi)^d} \frac{1}{(-k^2)^2} \bigg|_{UV} = \int \frac{d^d k}{(2\pi)^d} \frac{1}{(m^2 - k^2)^2} = \frac{im^{-2\varepsilon}}{(4\pi)^2} \Gamma(\varepsilon) = \frac{i}{(4\pi)^2} \frac{1}{\varepsilon} \qquad (1.30)$$

(see (1.2), (1.9)).

1.4 Feynman parametrization

Now we shall discuss a trick (invented by Feynman) to combine several denominators into a single one at the price of introducing some integrations. We'll not use it during these lectures; however, it is widely used, and it is important to know it. Let's multiply two α-parametrizations (1.5):

$$\frac{1}{a_1^{n_1} a_2^{n_2}} = \frac{1}{\Gamma(n_1)\Gamma(n_2)} \int e^{-a_1\alpha_1 - a_2\alpha_2} \alpha_1^{n_1-1} \alpha_2^{n_2-1} d\alpha_1 \, d\alpha_2 . \tag{1.31}$$

It is always possible to calculate one integral in such a product, namely, to integrate in a common scale η of all α_i. If we denote $\eta = \alpha_1 + \alpha_2$, i.e., make the substitution $\alpha_1 = \eta x$, $\alpha_2 = \eta(1-x)$, then we arrive at the Feynman parametrization

$$\frac{1}{a_1^{n_1} a_2^{n_2}} = \frac{\Gamma(n_1+n_2)}{\Gamma(n_1)\Gamma(n_2)} \int_0^1 \frac{x^{n_1-1}(1-x)^{n_2-1} dx}{[a_1 x + a_2(1-x)]^{n_1+n_2}} . \tag{1.32}$$

This is not the only possibility. We can also take, say, $\eta = \alpha_2$, i.e., make the substitution $\alpha_1 = \eta x$, $\alpha_2 = \eta$. This results in a variant of the Feynman parametrization, which can be more useful in some cases:

$$\frac{1}{a_1^{n_1} a_2^{n_2}} = \frac{\Gamma(n_1+n_2)}{\Gamma(n_1)\Gamma(n_2)} \int_0^\infty \frac{x^{n_1-1} dx}{[a_1 x + a_2]^{n_1+n_2}} . \tag{1.33}$$

Let's consider the general case of k denominators:

$$\frac{1}{a_1^{n_1} a_2^{n_2} \cdots a_k^{n_k}} = \frac{1}{\Gamma(n_1)\Gamma(n_2)\cdots\Gamma(n_k)}$$
$$\times \int e^{-a_1\alpha_1 - a_2\alpha_2 \cdots - a_k\alpha_k} \alpha_1^{n_1-1} \alpha_2^{n_2-1} \cdots \alpha_k^{n_k-1} d\alpha_1 \, d\alpha_2 \cdots d\alpha_k . \tag{1.34}$$

We can choose the common scale η to be the sum of any subset of α_i; the numbering of the denominators is not fixed, and we can always re-number them in such a way that $\eta = \alpha_1 + \alpha_2 \cdots + \alpha_l$, where $1 \le l \le k$. We insert

$$\delta(\alpha_1 + \alpha_2 \cdots + \alpha_l - \eta) d\eta$$

into the integrand, and make the substitution $\alpha_i = \eta x_i$. Then the integral in η can be easily calculated, and we obtain the general Feynman parametrization

$$\frac{1}{a_1^{n_1} a_2^{n_2} \cdots a_k^{n_k}} = \frac{\Gamma(n_1 + n_2 \cdots + n_k)}{\Gamma(n_1)\Gamma(n_2)\cdots\Gamma(n_k)}$$
$$\times \int \frac{\delta(x_1 + x_2 \cdots + x_l - 1)x_1^{n_1-1}x_2^{n_2-1}\cdots x_k^{n_k-1}dx_1\,dx_2\cdots dx_k}{[a_1 x_1 + a_2 x_2 \cdots + a_k x_k]^{n_1 + n_2 \cdots + n_k}}.$$

(1.35)

Let's stress once more that the δ function here can contain the sum of any subset of the variables x_i, from a single variable up to all of them.

1.5 Massless propagator diagram

Now we shall consider the massless propagator diagram (Fig. 1.6),

$$\int \frac{d^d k}{D_1^{n_1} D_2^{n_2}} = i\pi^{d/2}(-p^2)^{d/2-n_1-n_2}G(n_1, n_2),$$
$$D_1 = -(k+p)^2, \quad D_2 = -k^2$$

(1.36)

(from now on, we'll not write $-i0$ in denominators explicitly, but they are implied). We shall consider this diagram at $p^2 < 0$, i.e., below the threshold of production of a real pair, where the result is an analytic function. The power of $-p^2$ is evident from dimensionality. Our aim is to calculate the dimensionless function $G(n_1, n_2)$; we can put $-p^2 = 1$ to simplify things. This diagram is symmetric with respect to $1 \leftrightarrow 2$. It vanishes for integer $n_1 \leq 0$ or $n_2 \leq 0$, because then it becomes the massless vacuum diagram (Fig. 1.7), possibly, with some polynomial numerator.

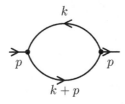

Fig. 1.6   One-loop massless propagator diagram

Using Wick rotation and α-parametrization (1.5), we rewrite the defi-

Fig. 1.7 Massless vacuum diagram

nition (1.36) of $G(n_1, n_2)$ as

$$G(n_1, n_2) = \frac{\pi^{-d/2}}{\Gamma(n_1)\Gamma(n_2)} \int e^{-\alpha_1(\boldsymbol{k}+\boldsymbol{p})^2 - \alpha_2 \boldsymbol{k}^2} \alpha_1^{n_1-1} \alpha_2^{n_2-1} d\alpha_1\, d\alpha_2\, d^d\boldsymbol{k}\,.$$
(1.37)

We want to separate a full square in the exponent; to this end, we shift the integration momentum:

$$\boldsymbol{k}' = \boldsymbol{k} + \frac{\alpha_1}{\alpha_1 + \alpha_2}\boldsymbol{p}\,,$$

and obtain

$$\begin{aligned}
G(n_1, n_2) &= \frac{\pi^{-d/2}}{\Gamma(n_1)\Gamma(n_2)} \int \exp\left[-\frac{\alpha_1\alpha_2}{\alpha_1 + \alpha_2}\right] \alpha_1^{n_1-1}\alpha_2^{n_2-1}d\alpha_1\, d\alpha_2 \\
&\quad \times \int e^{-(\alpha_1+\alpha_2)\boldsymbol{k}'^2} d^d\boldsymbol{k}' \\
&= \frac{1}{\Gamma(n_1)\Gamma(n_2)} \int \exp\left[-\frac{\alpha_1\alpha_2}{\alpha_1 + \alpha_2}\right] \frac{\alpha_1^{n_1-1}\alpha_2^{n_2-1}}{(\alpha_1 + \alpha_2)^{d/2}}d\alpha_1\, d\alpha_2\,.
\end{aligned}$$
(1.38)

Now we make the usual substitution $\alpha_1 = \eta x$, $\alpha_2 = \eta(1 - x)$, and obtain

$$\begin{aligned}
G(n_1, n_2) &= \frac{1}{\Gamma(n_1)\Gamma(n_2)} \int_0^1 x^{n_1-1}(1 - x)^{n_2-1}dx \\
&\quad \times \int_0^\infty e^{-\eta x(1-x)}\eta^{-d/2+n_1+n_2-1}d\eta \\
&= \frac{\Gamma(-d/2 + n_1 + n_2)}{\Gamma(n_1)\Gamma(n_2)} \int_0^1 x^{d/2-n_2-1}(1 - x)^{d/2-n_1-1}dx\,.
\end{aligned}$$
(1.39)

This integral is the Euler B-function, and we arrive at the final result

$$G(n_1, n_2) = \frac{\Gamma(-d/2 + n_1 + n_2)\Gamma(d/2 - n_1)\Gamma(d/2 - n_2)}{\Gamma(n_1)\Gamma(n_2)\Gamma(d - n_1 - n_2)}\,.$$
(1.40)

This function has a nice property

$$\frac{G(n_1, n_2 + 1)}{G(n_1, n_2)} = -\frac{(d - 2n_1 - 2n_2)(d - n_1 - n_2 - 1)}{n_2(d - 2n_2 - 2)} \tag{1.41}$$

which allows us to shift its arguments by ± 1. For all integer $n_{1,2}$, these integrals are proportional to

$$G_1 = G(1,1) = -\frac{2g_1}{(d-3)(d-4)}, \quad g_1 = \frac{\Gamma(1+\varepsilon)\Gamma^2(1-\varepsilon)}{\Gamma(1-2\varepsilon)}, \tag{1.42}$$

with coefficients which are rational functions of d.

The denominator in (1.36) behaves as $(k^2)^{n_1+n_2}$ at $k \to \infty$. Therefore, the integral diverges if $d \geq 2(n_1 + n_2)$. At $d \to 4$ this means $n_1 + n_2 \leq 2$. This ultraviolet divergence shows itself as a $1/\varepsilon$ pole of the first Γ function in the numerator of (1.40) for $n_1 = n_2 = 1$ (this Γ function depends on $n_1 + n_2$, i.e., on the behaviour of the integrand at $k \to \infty$). The integral (1.36) can also have infrared divergences. Its denominator behaves as $(k^2)^{n_2}$ at $k \to 0$, and the integral diverges in this region if $d \leq 2n_2$. At $d \to 4$ this means $n_2 \geq 2$. This infrared divergence shows itself as a $1/\varepsilon$ pole of the third Γ function in the numerator of (1.40) for $n_2 \geq 2$ (this Γ function depends on n_2, i.e., on the behaviour of the integrand at $k \to 0$). Similarly, the infrared divergence at $k + p \to 0$ appears, at $d \to 4$, as a pole of the second Γ function, if $n_1 \geq 2$.

Let's consider the integral (Fig. 1.6)

$$I(p^2) = -\frac{i}{\pi^{d/2}} \int \frac{d^d k}{(-k^2 - i0)(-(k+p)^2 - i0)} = G_1(-p^2)^{-\varepsilon} \tag{1.43}$$

as a function of the complex variable p^2. It has a cut along the positive half-axis $p^2 > 0$, starting at the branching point $p^2 = 0$ at the threshold of the real pair production (Fig. 1.8). Analytically continuing it from a negative $p^2 = -s$ (where the function is regular) along the upper contour in Fig. 1.8, $p^2 = -s \exp(-i\alpha)$ with α varying from 0 to π, we obtain

$$I(s + i0) = G_1 s^{-\varepsilon} e^{i\pi\varepsilon}. \tag{1.44}$$

Similarly, continuing along the lower contour, we get $I(s - i0)$ having the opposite sign of the imaginary part, and the discontinuity across the cut is

$$I(s + i0) - I(s - i0) = G_1 s^{-\varepsilon} 2i \sin(\pi\varepsilon). \tag{1.45}$$

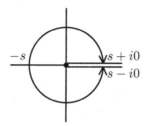

Fig. 1.8   Complex p^2 plane

At $\varepsilon \to 0$, G_1 (1.42) has a $1/\varepsilon$ pole, and the discontinuity is finite:

$$I(s+i0) - I(s-i0) = 2\pi i. \tag{1.46}$$

1.6   Tensors and γ-matrices in d dimensions

In order to calculate diagrams in d-dimensional space, we also need to do tensor and γ-matrix calculations in the numerators. For any integer d,

$$\delta_\mu^\mu = d \tag{1.47}$$

is the sum of d 1's. It is natural to extend this property to arbitrary d.

Let's consider the projector

$$\delta_{\nu_1}^{[\mu_1} \delta_{\nu_2}^{\mu_2} \cdots \delta_{\nu_n}^{\mu_n]} \tag{1.48}$$

onto completely antisymmetric tensors with n indices. Here the square brackets mean antisymmetrization, for example,

$$\delta_{\nu_1}^{[\mu_1} \delta_{\nu_2}^{\mu_2]} = \frac{1}{2!} \left(\delta_{\nu_1}^{\mu_1} \delta_{\nu_2}^{\mu_2} - \delta_{\nu_1}^{\mu_2} \delta_{\nu_2}^{\mu_1} \right),$$

$$\delta_{\nu_1}^{[\mu_1} \delta_{\nu_2}^{\mu_2} \delta_{\nu_3}^{\mu_3]} = \frac{1}{3!} \left(\delta_{\nu_1}^{\mu_1} \delta_{\nu_2}^{\mu_2} \delta_{\nu_3}^{\mu_3} - \delta_{\nu_1}^{\mu_1} \delta_{\nu_2}^{\mu_3} \delta_{\nu_3}^{\mu_2} + \delta_{\nu_1}^{\mu_2} \delta_{\nu_2}^{\mu_3} \delta_{\nu_3}^{\mu_1} - \delta_{\nu_1}^{\mu_2} \delta_{\nu_2}^{\mu_1} \delta_{\nu_3}^{\mu_3} \right.$$
$$\left. + \delta_{\nu_1}^{\mu_3} \delta_{\nu_2}^{\mu_1} \delta_{\nu_3}^{\mu_2} - \delta_{\nu_1}^{\mu_3} \delta_{\nu_2}^{\mu_2} \delta_{\nu_3}^{\mu_1} \right),$$

and so on. What is its trace

$$\delta_{\mu_1}^{[\mu_1} \delta_{\mu_2}^{\mu_2} \cdots \delta_{\mu_n}^{\mu_n]} ?$$

It is the number of independent components of an antisymmetric tensor with n indices in d-dimensional space. Let's consider an arbitrary integer d. All indices must be different; their order is not important. In other words, the indices μ_1, μ_2, $\ldots$, μ_n form a subset of the numbers 0, 1, $\ldots$, $(d-1)$. The number of such subsets, and hence the number of independent components of an antisymmetric tensor, is

$$\delta^{[\mu_1}_{\mu_1} \delta^{\mu_2}_{\mu_2} \cdots \delta^{\mu_n]}_{\mu_n} = \binom{d}{n} = \frac{1}{n!} d(d-1) \cdots (d-n+1) \,. \tag{1.49}$$

It is easy to check, for a few values of n, that this formula for arbitrary d follows from (1.47), for example,

$$\delta^{[\mu_1}_{\mu_1} \delta^{\mu_2]}_{\mu_2} = \frac{1}{2} \left(\delta^{\mu_1}_{\mu_1} \delta^{\mu_2}_{\mu_2} - \delta^{\mu_2}_{\mu_1} \delta^{\mu_1}_{\mu_2} \right) = \frac{1}{2} \left(d^2 - d \right) \,.$$

For an integer d, any tensor antisymmetric in $n > d$ indices is zero. In particular, the projectors (1.48) with $n > d$ vanish. Hence their traces vanish too, in accordance with (1.49). If d is not integer, the traces (1.49) are non-zero for all n. This means that the projectors (1.48) are non-zero. Therefore, in our d-dimensional space non-zero antisymmetric tensors with arbitrary numbers of indices exist. This space is, in fact, infinite-dimensional; it is the relation (1.47) which makes it look "d-dimensional".

Now we are going to discuss γ-matrices in d dimensions. Their defining property is

$$\gamma^\mu \gamma^\nu + \gamma^\nu \gamma^\mu = 2g^{\mu\nu} \,. \tag{1.50}$$

How many different products of the matrices γ^μ are there for an integer d? Any product of this kind can be simplified, using (1.50), to such a form that each of d matrices γ^μ occurs either 0 or 1 times. Therefore, the number of independent products is 2^d. For any even integer d, it is possible to define a set of Dirac matrices γ^μ in such a way that their products span the whole space of matrices. The number of independent $N \times N$ matrices is N^2. This means that γ^μ must be $2^{d/2} \times 2^{d/2}$ matrices, and the trace of the unit Dirac matrix is

$$\text{Tr}\,1 = 2^{d/2} \,. \tag{1.51}$$

For example, this is so in the familiar case $d = 4$.

In other words, any γ-matrix expression can be expanded in the basis

$$\Gamma^{\mu_1 \cdots \mu_n} = \gamma^{[\mu_1} \cdots \gamma^{\mu_n]} \,, \tag{1.52}$$

because symmetric parts are eliminated as a consequence of (1.50). For example,

$$\gamma^{\mu_1}\gamma^{\mu_2} = \gamma^{[\mu_1}\gamma^{\mu_2]} + g^{\mu_1\mu_2},$$
$$\gamma^{\mu_1}\gamma^{\mu_2}\gamma^{\mu_3} = \gamma^{[\mu_1}\gamma^{\mu_2}\gamma^{\mu_3]} + \gamma^{\mu_1}g^{\mu_2\mu_3} - \gamma^{\mu_2}g^{\mu_3\mu_1} + \gamma^{\mu_3}g^{\mu_1\mu_2},$$

and so on. For an integer d, this basis is finite: only antisymmetrized products with $n \le d$ exist. For a non-integer d, this basis is infinite.

It would be natural to accept (1.51) for any d. This formula can also be derived from the completeness relation of the basis (1.52) (this relation allows us to do Fierz transformations in d dimensions). However, the universally accepted convention is to use

$$\text{Tr}\,1 = 4 \tag{1.53}$$

instead (this somewhat simplifies formulas for diagrams with closed fermion loops). This convention can be understood in the following way. As we shall see, each closed quark loop in QCD produces the factor T_F (among other things). Changing the definition of the Dirac trace from (1.51) to (1.53) can be viewed as a d-dependent redefinition of T_F. In final renormalized results, the limit $d \to 4$ is taken, and T_F returns to its standard value (equal to $1/2$). The same argument applies to QED; the only difference is that the normal value of T_F (before the redefinition) in QED is 1, and it is usually not included into formulas explicitly. Because of (1.53), the usual formulas for traces of 2, 4,... γ-matrices hold.

Let's derive some relations which are used for simplifying γ-matrix expressions in d dimensions. From (1.50) we have

$$\gamma_\mu\gamma^\mu = d. \tag{1.54}$$

How to find $\gamma_\mu\slashed{a}\gamma^\mu$? We anticommute γ^μ to the left:

$$\gamma_\mu\slashed{a}\gamma^\mu = \gamma_\mu(-\gamma^\mu\slashed{a} + 2a^\mu) = -(d-2)\slashed{a}. \tag{1.55}$$

Similarly,

$$\gamma_\mu\slashed{a}\slashed{b}\gamma^\mu = \gamma_\mu\slashed{a}(-\gamma^\mu\slashed{b} + 2b^\mu) = (d-2)\slashed{a}\slashed{b} + 2\slashed{b}\slashed{a} = 4a\cdot b + (d-4)\slashed{a}\slashed{b}, \tag{1.56}$$

and

$$\begin{aligned}
\gamma_\mu\slashed{a}\slashed{b}\slashed{c}\gamma^\mu &= \gamma_\mu\slashed{a}\slashed{b}(-\gamma^\mu\slashed{c} + 2c^\mu) = -4a\cdot b\,\slashed{c} - (d-4)\slashed{a}\slashed{b}\slashed{c} + 2\slashed{c}\slashed{a}\slashed{b} \\
&= -2\slashed{c}\slashed{b}\slashed{a} - (d-4)\slashed{a}\slashed{b}\slashed{c}.
\end{aligned} \tag{1.57}$$

It is not possible to define a matrix γ_5 satisfying

$$\gamma_5\gamma^\mu + \gamma^\mu\gamma_5 = 0 \tag{1.58}$$

in d-dimensional space. Let us consider the following chain of equalities:

$$\operatorname{Tr}\gamma_5\gamma_\mu\gamma^\mu = d\operatorname{Tr}\gamma_5 = -\operatorname{Tr}\gamma_\mu\gamma_5\gamma^\mu = -\operatorname{Tr}\gamma_5\gamma^\mu\gamma_\mu = -d\operatorname{Tr}\gamma_5$$
$$\Rightarrow d\operatorname{Tr}\gamma_5 = 0$$

(the anticommutativity of γ_5 was used in the second step, and the trace cyclicity in the third one). We have learned that if $d \neq 0$ then $\operatorname{Tr}\gamma_5 = 0$. We assume that $d \neq 0$ and continue:

$$\operatorname{Tr}\gamma_5\gamma_\mu\gamma^\mu\gamma^\alpha\gamma^\beta = d\operatorname{Tr}\gamma_5\gamma^\alpha\gamma^\beta = -\operatorname{Tr}\gamma_5\gamma_\mu\gamma^\alpha\gamma^\beta\gamma^\mu$$
$$= -(d-4)\operatorname{Tr}\gamma_5\gamma^\alpha\gamma^\beta \Rightarrow (d-2)\operatorname{Tr}\gamma_5\gamma^\alpha\gamma^\beta = 0\,.$$

We have learned that if $d \neq 2$ then $\operatorname{Tr}\gamma_5\gamma^\alpha\gamma^\beta = 0$. We assume that $d \neq 2$ and continue:

$$\operatorname{Tr}\gamma_5\gamma_\mu\gamma^\mu\gamma^\alpha\gamma^\beta\gamma^\gamma\gamma^\delta = d\operatorname{Tr}\gamma_5\gamma^\alpha\gamma^\beta\gamma^\gamma\gamma^\delta = -\operatorname{Tr}\gamma_5\gamma_\mu\gamma^\alpha\gamma^\beta\gamma^\gamma\gamma^\delta\gamma^\mu$$
$$= -(d-8)\operatorname{Tr}\gamma_5\gamma^\alpha\gamma^\beta\gamma^\gamma\gamma^\delta \Rightarrow (d-4)\operatorname{Tr}\gamma_5\gamma^\alpha\gamma^\beta\gamma^\gamma\gamma^\delta = 0\,,$$

where we have used $\gamma_\mu\gamma^\alpha\gamma^\beta\gamma^\gamma\gamma^\delta\gamma^\mu = (d-8)\gamma^\alpha\gamma^\beta\gamma^\gamma\gamma^\delta$+terms with fewer γ-matrices. We have learned that if $d \neq 4$ then $\operatorname{Tr}\gamma_5\gamma^\alpha\gamma^\beta\gamma^\gamma\gamma^\delta = 0$. Assuming $d \neq 4$, we can show that $(d-6)\operatorname{Tr}\gamma_5\gamma^\alpha\gamma^\beta\gamma^\gamma\gamma^\delta\gamma^\varepsilon\gamma^\zeta = 0$, and so on. All traces vanish if d is not an even integer. And this is not what we want for γ_5.

This leads to interesting physical consequences (the axial anomaly). In QED and QCD, γ_5 does not appear in the Lagrangian, and hence in calculations of S-matrix elements. Difficulties only appear when we consider some external operators, such as axial quark currents. The problem becomes much more severe in the electroweak theory, where γ_5 appears in the Lagrangian (and still more severe in supersymmetric theories).

Chapter 2

QED at one loop

2.1 Lagrangian and Feynman rules

First we shall discuss quantum electrodynamics with massless electron. This theory has many similarities with QCD. Its Lagrangian is

$$L = \bar{\psi}_0 i \slashed{D} \psi_0 - \frac{1}{4} F_{0\mu\nu} F_0^{\mu\nu} \,, \qquad (2.1)$$

where ψ_0 is the electron field,

$$D_\mu \psi_0 = (\partial_\mu - ie_0 A_{0\mu}) \psi_0 \qquad (2.2)$$

is its covariant derivative, $A_{0\mu}$ is the photon field, and

$$F_{0\mu\nu} = \partial_\mu A_{0\nu} - \partial_\nu A_{0\mu} \qquad (2.3)$$

is the field strength tensor.

As explained in the textbooks, it is not possible to obtain the photon propagator from this Lagrangian: the matrix which should be inverted has zero determinant. This is due to the gauge invariance; in order to have a photon propagator, we have to fix the gauge in some way. The most popular way (called the covariant gauge) is to add the term

$$\Delta L = -\frac{1}{2a_0} \left(\partial_\mu A_0^\mu \right)^2 \qquad (2.4)$$

to the Lagrangian (2.1) (a_0 is the gauge-fixing parameter). This additional term does not change physics: physical results are gauge-invariant, and do not depend on the choice of the gauge (in particular, on a_0).

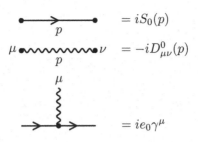

Fig. 2.1   QED Feynman rules

Now we can derive the Feynman rules (Fig. 2.1). The free electron propagator is

$$S_0(p) = \frac{1}{\not{p}} = \frac{\not{p}}{p^2}, \tag{2.5}$$

and the free photon propagator is

$$D^0_{\mu\nu}(p) = \frac{1}{p^2}\left[g_{\mu\nu} - (1 - a_0)\frac{p_\mu p_\nu}{p^2} \right]. \tag{2.6}$$

The Lagrangian (2.1), (2.4) contains the bare fields ψ_0, A_0 and the bare parameters e_0, a_0. The renormalized quantities are related to them by

$$\psi_0 = Z_\psi^{1/2}\psi, \quad A_0 = Z_A^{1/2}A, \quad a_0 = Z_A a, \quad e_0 = Z_\alpha^{1/2}e \tag{2.7}$$

(we shall see why the same renormalization constant Z_A describes renormalization of both the photon field A and the gauge-fixing parameter a in Sect. 2.2). We shall mostly use the *minimal renormalization scheme* ($\overline{\text{MS}}$). In this scheme, renormalization constants have the structure

$$Z_i(\alpha) = 1 + \frac{z_1}{\varepsilon}\frac{\alpha}{4\pi} + \left(\frac{z_{22}}{\varepsilon^2} + \frac{z_{21}}{\varepsilon}\right)\left(\frac{\alpha}{4\pi}\right)^2 + \cdots \tag{2.8}$$

They contain neither finite parts (ε^0) nor positive powers of ε, just $1/\varepsilon^n$ terms necessary to remove $1/\varepsilon^n$ divergences in renormalized (physical) results. Sometimes, other renormalization schemes are useful, see Chap. 5.

In order to write renormalization constants (2.8), we must define the renormalized coupling α in such a way that it is exactly dimensionless at any d. The action is dimensionless, because it appears in the exponent in the Feynman path integral. The action is an integral of L over d-dimensional space–time; therefore, the dimensionality of the Lagrangian L is $[L] = d$ (in mass units). From the kinetic terms in the Lagrangian (2.1) we obtain the dimensionalities of the fields: $[A_0] = 1 - \varepsilon$, $[\psi_0] = 3/2 - \varepsilon$. Both terms in the covariant derivative (2.2) must have the same dimensionality 1, therefore $[e_0] = \varepsilon$. In order to define a dimensionless coupling α, we have to introduce a parameter μ with the dimensionality of mass (called the renormalization scale):

$$\frac{\alpha(\mu)}{4\pi} = \mu^{-2\varepsilon} \frac{e^2}{(4\pi)^{d/2}} e^{-\gamma\varepsilon}, \qquad (2.9)$$

where γ is the Euler constant[1]. In practise, this equation is more often used in the opposite direction:

$$\frac{e_0^2}{(4\pi)^{d/2}} = \mu^{2\varepsilon} \frac{\alpha(\mu)}{4\pi} Z_\alpha(\alpha(\mu)) e^{\gamma\varepsilon}. \qquad (2.10)$$

We first calculate some physical quantity in terms of the bare charge e_0, and then re-express it via the renormalized $\alpha(\mu)$.

2.2   Photon propagator

The photon propagator has the structure (Fig. 2.2)

$$\begin{aligned} -iD_{\mu\nu}(p) = {}& -iD^0_{\mu\nu}(p) + (-i)D^0_{\mu\alpha}(p)i\Pi^{\alpha\beta}(p)(-i)D^0_{\beta\nu}(p) \\ & + (-i)D^0_{\mu\alpha}(p)i\Pi^{\alpha\beta}(p)(-i)D^0_{\beta\gamma}(p)i\Pi^{\gamma\delta}(p)(-i)D^0_{\gamma\nu}(p) + \cdots \end{aligned} \qquad (2.11)$$

where the photon self-energy $i\Pi_{\mu\nu}(p)$ (denoted by a shaded blob in Fig. 2.2) is the sum of all one-particle-irreducible diagrams (diagrams which cannot be cut into two disconnected pieces by cutting a single photon line), not including the external photon propagators.

The series (2.11) can also be rewritten as an equation

$$D_{\mu\nu}(p) = D^0_{\mu\nu}(p) + D^0_{\mu\alpha}(p)\Pi^{\alpha\beta}(p)D_{\beta\nu}(p). \qquad (2.12)$$

_____

[1] The first renormalization scheme in the framework of dimensional regularization was called MS (minimal subtractions); in this scheme

$$\alpha(\mu) = \mu^{-2\varepsilon} \frac{e^2}{4\pi}.$$

It soon became clear that results for loop diagrams in this scheme look unnecessarily complicated, and the $\overline{\text{MS}}$ (modified minimal subtractions) scheme (2.9) was proposed. Some authors use slightly different definitions of the $\overline{\text{MS}}$ scheme, with $\Gamma(1+\varepsilon)$ or $1/\Gamma(1-\varepsilon)$ instead of $e^{-\gamma\varepsilon}$ in (2.9).

Fig. 2.2   Photon propagator

In order to solve this equation, let's introduce, for any tensor of the form

$$A_{\mu\nu} = A_\perp \left[g_{\mu\nu} - \frac{p_\mu p_\nu}{p^2} \right] + A_\| \frac{p_\mu p_\nu}{p^2} \,,$$

the inverse tensor

$$A_{\mu\nu}^{-1} = A_\perp^{-1} \left[g_{\mu\nu} - \frac{p_\mu p_\nu}{p^2} \right] + A_\|^{-1} \frac{p_\mu p_\nu}{p^2}$$

satisfying

$$A_{\mu\lambda}^{-1} A^{\lambda\nu} = \delta_\mu^\nu \,.$$

Then

$$D_{\mu\nu}^{-1}(p) = (D^0)_{\mu\nu}^{-1}(p) - \Pi_{\mu\nu}(p) \,. \tag{2.13}$$

As we shall see in Sect. 2.3,

$$\Pi_{\mu\nu}(p) p^\nu = 0 \,, \quad \Pi_{\mu\nu}(p) p^\mu = 0 \,. \tag{2.14}$$

Therefore, the photon self-energy has the form

$$\Pi_{\mu\nu}(p) = (p^2 g_{\mu\nu} - p_\mu p_\nu) \Pi(p^2) \,, \tag{2.15}$$

and the full photon propagator is

$$D_{\mu\nu}(p) = \frac{1}{p^2(1 - \Pi(p^2))} \left[g_{\mu\nu} - \frac{p_\mu p_\nu}{p^2} \right] + a_0 \frac{p_\mu p_\nu}{(p^2)^2} \,. \tag{2.16}$$

Its longitudinal part gets no corrections, to all orders of perturbation theory. The full bare propagator is related to the renormalized one by

$$D_{\mu\nu}(p) = Z_A(\alpha(\mu)) D_{\mu\nu}^r(p; \mu) \,.$$

Therefore,

$$D_{\mu\nu}^r(p; \mu) = D_\perp^r(p^2; \mu) \left[g_{\mu\nu} - \frac{p_\mu p_\nu}{p^2} \right] + a(\mu) \frac{p_\mu p_\nu}{(p^2)^2} \,. \tag{2.17}$$

The minimal (2.8) renormalization constant $Z_A(\alpha)$ is constructed to make

$$D_\perp^r(p^2; \mu) = Z_A^{-1}(\alpha(\mu)) \frac{1}{p^2(1 - \Pi(p^2))}$$

finite at $\varepsilon \to 0$. But the longitudinal part of (2.17) containing

$$a(\mu) = Z_A^{-1}(\alpha(\mu)) a_0$$

must be finite too. This explains why Z_A appears twice in (2.7), in renormalization of A_0 and of a_0.

2.3 Ward identity

It is easy to check by a direct calculation that (Fig. 2.3)

$$iS_0(p')\, ie_0 \slashed{q}\, iS_0(p) = e_0 \left[iS_0(p') - iS_0(p)\right]. \tag{2.18}$$

Here $q = p' - p$; substituting $\slashed{q}$ and using $S_0(p) = 1/\slashed{p}$, we immediately obtain (2.18). We shall use graphical notation (Fig. 2.3): a photon line with a black triangle at the end means an external gluon leg (no propagator!) contracted in its polarization index with the incoming photon momentum q. A dot near an electron line means that its momentum is shifted by q, as compared to the diagram without the longitudinal photon insertion. For an infinitesimal q, we obtain from (2.18) a useful identity

$$\frac{\partial S_0(p)}{\partial p^\mu} = -S_0(p)\gamma_\mu S_0(p). \tag{2.19}$$

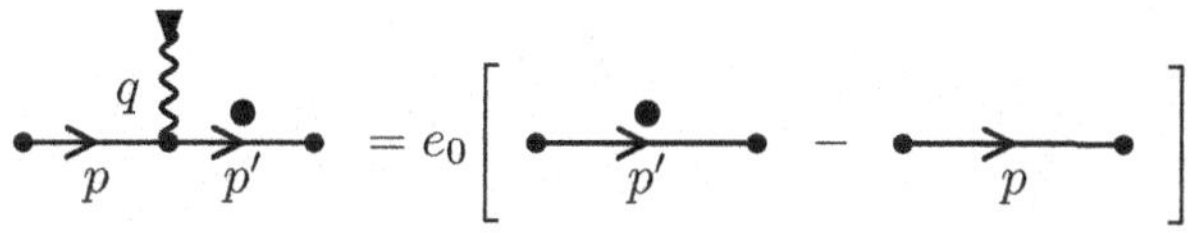

Fig. 2.3   Ward identity

Let's calculate $\Pi_{\mu\nu}(p)p^\nu$ (2.14) at one loop:

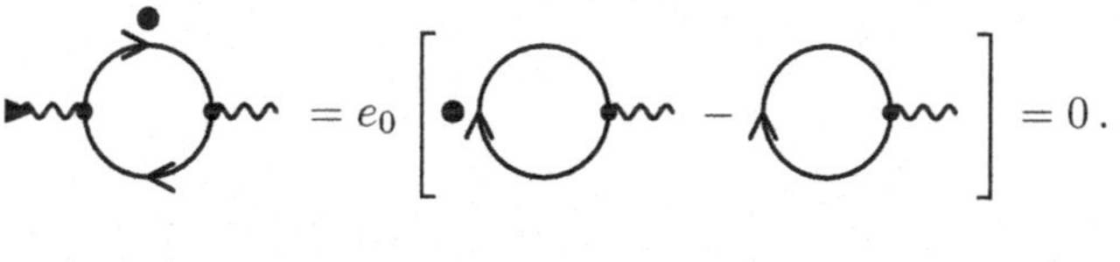

Two integrals here differ by a shift of the integration momentum, and hence
are equal.

Now we'll do the same at two loops. All two-loop diagrams for $\Pi_{\mu\nu}(p)p^\nu$
can be obtained from one two-loop diagram with a single photon leg, and
using the Ward identity (Fig. 2.3) we obtain

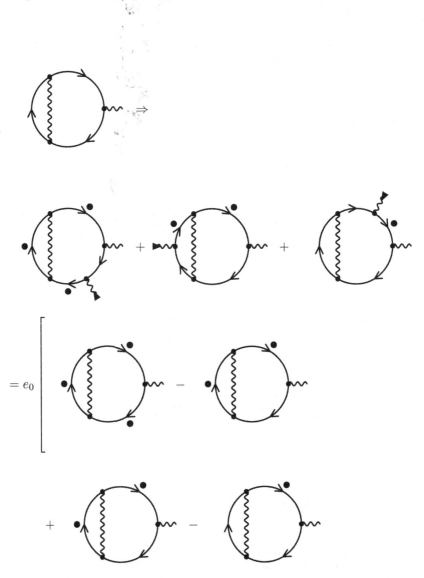

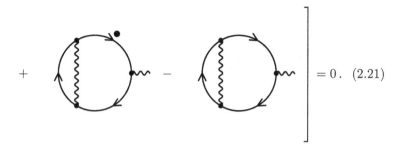

$$+ \qquad - \qquad \Bigg] = 0 \,. \quad (2.21)$$

All diagrams cancel pairwise, except the first one and the last one; these two diagrams differ only by a shift in an integration momentum.

It is clear that this proof works at any order of perturbation theory.

2.4 Photon self-energy

Now we shall explicitly calculate photon self-energy at one loop (Fig. 2.4). The fermion loop gives the factor -1, and

$$i(p^2 g_{\mu\nu} - p_\mu p_\nu)\Pi(p^2) = - \int \frac{d^d k}{(2\pi)^d} \, \mathrm{Tr} \, ie_0\gamma_\mu i\frac{\slashed{k}+\slashed{p}}{(k+p)^2} ie_0\gamma_\nu \frac{\slashed{k}}{k^2} \,. \quad (2.22)$$

To simplify finding the scalar function $\Pi(p^2)$, we contract in μ and ν. We obtain (see (1.47))

$$\Pi(p^2) = \frac{-ie_0^2}{(d-1)(-p^2)} \int \frac{d^d k}{(2\pi)^d} \frac{\mathrm{Tr}\,\gamma_\mu(\slashed{k}+\slashed{p})\gamma^\mu\slashed{k}}{[-(k+p)^2]\,(-k^2)} \,. \quad (2.23)$$

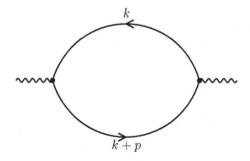

Fig. 2.4 One-loop photon self-energy

The trace in the numerator of (2.23) can be found using (1.55):

$$\Pi(p^2) = \frac{d-2}{d-1}\frac{ie_0^2}{-p^2}\int\frac{d^d k}{(2\pi)^d}\frac{4(k+p)\cdot k}{[-(k+p)^2](-k^2)}\,. \tag{2.24}$$

We can set $-p^2 = 1$; it is easy to restore the power of $-p^2$ by dimensionality at the end of calculation. The denominators (1.36) are

$$D_1 = -(k+p)^2\,,\quad D_2 = -k^2\,.$$

All scalar products in the numerator can be expressed via the denominators:

$$p^2 = -1\,,\quad k^2 = -D_2\,,\quad p\cdot k = \frac{1}{2}\left(1 + D_2 - D_1\right)\,. \tag{2.25}$$

We obtain

$$\Pi(p^2) = 2\frac{d-2}{d-1}ie_0^2\int\frac{d^d k}{(2\pi)^d}\frac{-2D_2 + 1 + D_2 - D_1}{D_1 D_2}\,. \tag{2.26}$$

Terms with D_1 or D_2 in the numerator can be omitted, because integrals with a single massless denominator (Fig. 1.7) vanish.

Restoring the power of $-p^2$, we arrive at the final result

$$\Pi(p^2) = -\frac{e_0^2(-p^2)^{-\varepsilon}}{(4\pi)^{d/2}}2\frac{d-2}{d-1}G_1\,, \tag{2.27}$$

or, recalling (1.42),

$$\Pi(p^2) = \frac{e_0^2(-p^2)^{-\varepsilon}}{(4\pi)^{d/2}}4\frac{d-2}{(d-1)(d-3)(d-4)}g_1\,. \tag{2.28}$$

2.5   Photon field renormalization

The transverse part of the photon propagator (2.16) is, with the one-loop accuracy,

$$p^2 D_\perp(p^2) = \frac{1}{1-\Pi(p^2)} = 1 + \frac{e_0^2(-p^2)^{-\varepsilon}}{(4\pi)^{d/2}}4\frac{d-2}{(d-1)(d-3)(d-4)}g_1\,. \tag{2.29}$$

Re-expressing it via the renormalized $\alpha(\mu)$ (2.10), we obtain

$$p^2 D_\perp(p^2) = 1 + \frac{\alpha(\mu)}{4\pi} e^{-L\varepsilon} e^{\gamma\varepsilon} g_1 \, 4 \frac{d-2}{(d-1)(d-3)(d-4)} \,,$$

$$L = \log \frac{-p^2}{\mu^2} \,. \tag{2.30}$$

We want to expand this at $\varepsilon \to 0$. Using (1.12), we see that

$$e^{\gamma\varepsilon} g_1 = 1 + \mathcal{O}(\varepsilon^2) \,.$$

This is exactly the reason of including $\exp(-\gamma\varepsilon)$ into the definition (2.9). We obtain

$$p^2 D_\perp(p^2) = 1 - \frac{4}{3} \frac{\alpha(\mu)}{4\pi\varepsilon} \left[1 - \left(L - \frac{5}{3} \right) \varepsilon + \cdots \right] \,. \tag{2.31}$$

This should be equal to $Z_A(\alpha(\mu)) p^2 D_\perp^r(p^2; \mu)$, where $Z_A(\alpha)$ is a minimal (2.8) renormalization constant, and $D_\perp^r(p^2; \mu)$ is finite at $\varepsilon \to 0$. It is easy to see that

$$Z_A(\alpha) = 1 - \frac{4}{3} \frac{\alpha}{4\pi\varepsilon} \,, \tag{2.32}$$

and

$$p^2 D_\perp^r(p^2; \mu) = 1 + \frac{4}{3} \frac{\alpha(\mu)}{4\pi} \left(L - \frac{5}{3} \right) \,. \tag{2.33}$$

The bare propagator $D_\perp(p^2) = Z_A(\alpha(\mu)) D_\perp^r(p^2; \mu)$ does not depend on μ. Differentiating it in $\log\mu$ we obtain the renormalization group (RG) equation

$$\frac{\partial D_\perp^r(p^2; \mu)}{\partial \log\mu} + \gamma_A(\alpha(\mu)) D_\perp^r(p^2; \mu) = 0 \,, \tag{2.34}$$

where the anomalous dimension is defined by

$$\gamma_A(\alpha(\mu)) = \frac{d\log Z_A(\alpha(\mu))}{d\log\mu} \,. \tag{2.35}$$

For any minimal renormalization constant

$$Z_i(\alpha) = 1 + z_1 \frac{\alpha}{4\pi\varepsilon} + \cdots$$

using (2.9)

$$\frac{d\log\alpha(\mu)}{d\log\mu} = -2\varepsilon + \cdots$$

we find the corresponding anomalous dimension

$$\gamma_i(\alpha(\mu)) = \frac{d\log Z_i(\alpha(\mu))}{d\log\mu} = \gamma_0 \frac{\alpha(\mu)}{4\pi} + \cdots$$

to be

$$\gamma_i(\alpha) = -2z_1 \frac{\alpha}{4\pi} + \cdots$$

In other words, the renormalization constant with the one-loop accuracy is

$$Z_i(\alpha) = 1 - \frac{\gamma_0}{2}\frac{\alpha}{4\pi\varepsilon} + \cdots$$

The anomalous dimension of the photon field is, from (2.32),

$$\gamma_A(\alpha) = \frac{8}{3}\frac{\alpha}{4\pi} + \cdots \tag{2.36}$$

The dependence of the renormalized propagator (2.33) on L can be completely determined from the RG equation. We can rewrite it as

$$\frac{\partial p^2 D_\perp^r}{\partial L} = \frac{\gamma_A}{2} p^2 D_\perp^r. \tag{2.37}$$

Solving this equation with the initial condition

$$p^2 D_\perp^r(p^2; \mu^2 = -p^2) = 1 - \frac{20}{9}\frac{\alpha(\mu)}{4\pi}$$

at $L = 0$, we can reconstruct (2.33).

We can also write the RG equation for $a(\mu)$. The bare gauge-fixing parameter $a_0 = Z_A(\alpha(\mu))a(\mu)$ does not depend on μ. Differentiating it in $\log\mu$, we obtain

$$\frac{da(\mu)}{d\log\mu} + \gamma_A(\alpha(\mu))a(\mu) = 0. \tag{2.38}$$

2.6   Electron propagator

The electron propagator has the structure (Fig. 2.5)

$$iS(p) = iS_0(p) + iS_0(p)(-i)\Sigma(p)iS_0(p) \\ + iS_0(p)(-i)\Sigma(p)iS_0(p)(-i)\Sigma(p)iS_0(p) + \cdots \tag{2.39}$$

where the electron self-energy $-i\Sigma(p)$ is the sum of all one-particle-irreducible diagrams (diagrams which cannot be cut into two disconnected pieces by cutting a single electron line), not including the external electron propagators.

Fig. 2.5   Electron propagator

The series (2.39) can also be rewritten as an equation

$$S(p) = S_0(p) + S_0(p)\Sigma(p)S(p). \tag{2.40}$$

Its solution is

$$S(p) = \frac{1}{S_0^{-1}(p) - \Sigma(p)}. \tag{2.41}$$

Electron self-energy $\Sigma(p)$ depends on a single vector p, and can have two γ-matrix structures: 1 and $\not{p}$. When electron is massless, any diagram for Σ contains an odd number of γ matrices, and the structure 1 cannot appear:

$$\Sigma(p) = \not{p}\Sigma_V(p^2). \tag{2.42}$$

This is due to helicity conservation. In massless QED, the electrons with helicity $\lambda = \mp\frac{1}{2}$,

$$\psi_{L,R} = \frac{1 \pm \gamma_5}{2}\psi,$$

cannot transform into each other. Operators with an odd number of γ matrices, like (2.42), conserve helicity, and those with an even number of γ matrices flip helicity. Therefore, the massless electron propagator has the form

$$S(p) = \frac{1}{1 - \Sigma_V(p^2)}\frac{1}{\not{p}}. \tag{2.43}$$

Let's calculate electron self-energy at one loop (Fig. 2.6):

$$-i\not{p}\Sigma_V(p^2) = \int \frac{d^d k}{(2\pi)^d} ie_0\gamma^\mu i\frac{\not{k}+\not{p}}{(k+p)^2} ie_0\gamma^\nu \frac{-i}{k^2}\left(g_{\mu\nu} - \xi\frac{k_\mu k_\nu}{k^2}\right), \tag{2.44}$$

where we introduced the notation

$$\xi = 1 - a_0. \tag{2.45}$$

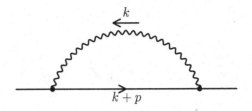

Fig. 2.6   One-loop electron self-energy

To find the scalar function $\Sigma_V(p^2)$, we take $\frac{1}{4}\operatorname{Tr}\not{p}$ of both sides:

$$\Sigma_V(p^2) = \frac{ie_0^2}{-p^2}\int\frac{d^dk}{(2\pi)^d}\frac{N}{D_1D_2},$$

$$N = \frac{1}{4}\operatorname{Tr}\not{p}\gamma^\mu(\not{k}+\not{p})\gamma^\nu\left(g_{\mu\nu}+\xi\frac{k_\mu k_\nu}{D_2}\right). \tag{2.46}$$

Using the "multiplication table" (2.25) we obtain

$$N = \frac{1}{4}\operatorname{Tr}\not{p}\gamma_\mu(\not{k}+\not{p})\gamma^\mu + \frac{\xi}{D_2}\frac{1}{4}\operatorname{Tr}\not{p}\not{k}(\not{k}+\not{p})\not{k}$$

$$= -(d-2)(p^2+p\cdot k) + \frac{\xi}{D_2}\left[k^2 p\cdot k + 2(p\cdot k)^2 - p^2 k^2\right] \tag{2.47}$$

$$\Rightarrow \frac{1}{2}\left[d-2+\xi\left(\frac{1}{D_2}-1\right)\right],$$

where terms with D_1 or D_2 in the numerator were omitted. Restoring the power of $-p^2$, we have

$$\Sigma_V(p^2) = -\frac{e_0^2(-p^2)^{-\varepsilon}}{(4\pi)^{d/2}}\frac{1}{2}\left[(d-2-\xi)G(1,1)+\xi G(1,2)\right]. \tag{2.48}$$

Using (1.41), we have

$$\frac{G(1,2)}{G(1,1)} = -(d-3), \tag{2.49}$$

and we arrive at

$$\Sigma_V(p^2) = -\frac{e_0^2(-p^2)^{-\varepsilon}}{(4\pi)^{d/2}}\frac{d-2}{2}a_0 G_1. \tag{2.50}$$

Note that the one-loop electron self-energy vanishes in the Landau gauge

$a_0 = 0$. Recalling (1.42), we can rewrite the result as

$$\Sigma_V(p^2) = \frac{e_0^2(-p^2)^{-\varepsilon}}{(4\pi)^{d/2}} \frac{d-2}{(d-3)(d-4)} a_0 g_1 \, . \tag{2.51}$$

The electron propagator (2.43), with one-loop accuracy, expressed via the renormalized quantities $\alpha(\mu)$ (2.10) and $a(\mu)$ (2.7) (we may take $a(\mu) = a_0$, because it only appears in the α correction) is

$$\begin{aligned}
\not{p}S(p) &= 1 + \frac{\alpha(\mu)}{4\pi} e^{-L\varepsilon} e^{\gamma\varepsilon} g_1 \, a(\mu) \frac{d-2}{(d-3)(d-4)} \\
&= 1 - \frac{\alpha(\mu)}{4\pi\varepsilon} a(\mu) e^{-L\varepsilon} (1 + \varepsilon + \cdots) \, .
\end{aligned} \tag{2.52}$$

It should be equal to $Z_\psi(\alpha(\mu), a(\mu)) \not{p} S_r(p; \mu)$, where $Z_\psi(\alpha)$ is a minimal (2.8) renormalization constant, and $S_r(p; \mu)$ is finite at $\varepsilon \to 0$. It is easy to see that

$$Z_\psi(\alpha, a) = 1 - a\frac{\alpha}{4\pi\varepsilon} \tag{2.53}$$

and

$$\not{p}S_r(p; \mu) = 1 + a(\mu)(L-1)\frac{\alpha(\mu)}{4\pi} \, . \tag{2.54}$$

The bare propagator $S(p) = Z_\psi(\alpha(\mu))S_r(p; \mu)$ does not depend on μ. Differentiating it in $\log\mu$ we obtain the RG equation

$$\frac{\partial S_r(p; \mu)}{\partial \log \mu} + \gamma_\psi(\alpha(\mu), a(\mu)) S_r(p; \mu) = 0 \, , \tag{2.55}$$

where the anomalous dimension

$$\gamma_\psi(\alpha(\mu), a(\mu)) = \frac{d\log Z_\psi(\alpha(\mu), a(\mu))}{d\log\mu} \tag{2.56}$$

is

$$\gamma_\psi(\alpha, a) = 2a\frac{\alpha}{4\pi} + \cdots \tag{2.57}$$

2.7 Vertex and charge renormalization

Let the sum of all one-particle-irreducible vertex diagrams, i.e. diagrams which cannot be cut into disconnected pieces by cutting a single electron or photon line, not including the external propagators, be the vertex

$ie_0\Gamma^\mu(p,p')$ (Fig. 2.7). It can be written as

$$\Gamma^\mu(p,p') = \gamma^\mu + \Lambda^\mu(p,p'),\tag{2.58}$$

where $\Lambda^\mu(p,p')$ starts from one loop.

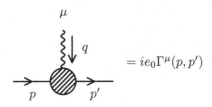

Fig. 2.7　Electron–photon vertex

When expressed via renormalized quantities, the vertex should be equal to

$$\Gamma^\mu = Z_\Gamma \Gamma_r^\mu,\tag{2.59}$$

where Z_Γ is a minimal (2.8) renormalization constant, and the renormalized vertex Γ_r^μ is finite at $\varepsilon \to 0$.

In order to obtain a physical scattering amplitude (S-matrix element), one should calculate the corresponding vertex (one-particle-irreducible, external propagators are not included) and multiply it by the spin wave functions of the external particles and by the field renormalization constants $Z_i^{1/2}$ for each external particle i. We can understand this rule in the following way. In fact, there are no external legs, only propagators. Suppose we are studying photon scattering in the laboratory. This photon has been emitted somewhere. Even if it was emitted in a far star (Fig. 2.8), there is a photon propagator from the far star to the laboratory. The (bare) propagator contains the factor Z_A. We split it into $Z_A^{1/2} \cdot Z_A^{1/2}$, and put one factor $Z_A^{1/2}$ into the emission process in the far star, and the other factor $Z_A^{1/2}$ into the scattering process in the laboratory.

The physical matrix element $e_0\Gamma Z_\psi Z_A^{1/2}$ of photon emission (or absorption) by an electron must be finite at $\varepsilon \to 0$. Strictly speaking, it is not an S-matrix element, because at least one particle must be off-shell, but this does not matter. This matrix element can be rewritten as $e\Gamma_r Z_\alpha^{1/2} Z_\Gamma Z_\psi Z_A^{1/2}$. The renormalized charge e and the renormalized vertex Γ_r are finite. Therefore, the minimal (2.8) renormalization constant $Z_\alpha^{1/2} Z_\Gamma Z_\psi Z_A^{1/2}$ must be

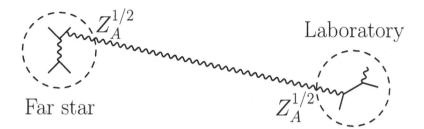

Fig. 2.8   Scattering of a photon emitted in a far star

finite, too. According to the definition, the only minimal renormalization constant which is finite at $\varepsilon \to 0$ is 1. Therefore,

$$Z_\alpha = (Z_\Gamma Z_\psi)^{-2} Z_A^{-1}. \tag{2.60}$$

In order to obtain the charge renormalization constant Z_α, one has to find the vertex renormalization constant Z_Γ and the electron- and photon-field renormalization constants Z_ψ and Z_A. In fact, the situation in QED is simpler, because $Z_\Gamma Z_\psi = 1$, due to the Ward identity (Sect. 2.3).

Starting from each diagram for $-i\Sigma(p)$, we can construct a set of diagrams for $ie_0 \Lambda^\mu(p, p')$, by attaching the external photon line to each electron propagator in turn. Let's calculate the contribution of this set to $ie_0 \Lambda^\mu(p, p') q_\mu$ using the Ward identity of Fig. 2.3. As an example, we consider all vertex diagrams generated from a certain two-loop electron self-energy diagram:

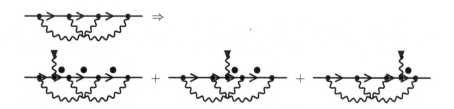

$$(2.61)$$

Of course, such cancellations happen for vertex diagrams generated from any self-energy diagram (and attaching the external photon line to an electron loop gives 0, due to C-parity conservation). Therefore, we arrive at the Ward–Takahashi identity

$$\Lambda^\mu(p, p')q_\mu = \Sigma(p) - \Sigma(p'),$$ (2.62)

which can also be rewritten as

$$\Gamma^\mu(p, p')q_\mu = S^{-1}(p') - S^{-1}(p).$$ (2.63)

For $q \to 0$, we obtain the Ward identity

$$\Lambda^\mu(p, p) = -\frac{\partial \Sigma(p)}{\partial p_\mu} \quad \text{or} \quad \Gamma^\mu(p, p) = \frac{\partial S^{-1}(p)}{\partial p_\mu}.$$ (2.64)

Rewriting (2.63) via the renormalized quantities,

$$Z_\Gamma \Gamma_r^\mu q_\mu = Z_\psi^{-1} \left[S_r^{-1}(p') - S_r^{-1}(p) \right],$$

we see that $Z_\psi Z_\Gamma$ must be finite. But the only minimal renormalization constant finite at $\varepsilon \to 0$ is 1:

$$Z_\psi Z_\Gamma = 1.$$ (2.65)

Therefore, charge renormalization in QED is determined by the photon field renormalization:

$$Z_\alpha = Z_A^{-1}.$$ (2.66)

We know Z_Γ at one loop from the Ward identity (2.65) and Z_ψ (2.53). Nevertheless, let's also find it by a direct calculation. This will be useful, because we'll have to do several similar calculations in QCD. We are only interested in the ultraviolet divergence of the diagram in Fig. 2.9. This divergence is logarithmic. We may nullify all external momenta, because terms which depend on these momenta are convergent:

$$ie_0\Lambda^\alpha = \int \frac{d^dk}{(2\pi)^d} ie_0\gamma^\mu i\frac{\not{k}}{k^2} ie_0\gamma^\alpha i\frac{\not{k}}{k^2} ie_0\gamma^\nu \frac{-i}{k^2} \left(g_{\mu\nu} - \xi\frac{k_\mu k_\nu}{k^2} \right) . \tag{2.67}$$

Of course, we should introduce some infrared regularization, otherwise this diagram vanishes. We have

$$\Lambda^\alpha = -ie_0^2 \int \frac{d^dk}{(2\pi)^d} \frac{\gamma_\mu \not{k}\gamma^\alpha \not{k}\gamma^\mu - \xi k^2\gamma^\alpha}{(k^2)^2} . \tag{2.68}$$

Averaging over k directions:

$$\not{k}\gamma^\alpha \not{k} \rightarrow \frac{k^2}{d}\gamma_\nu \gamma^\alpha \gamma^\nu ,$$

we obtain (4-dimensional γ-matrix algebra may be used)

$$\Lambda^\alpha = -ie_0^2 a_0 \gamma^\alpha \int \frac{d^dk}{(2\pi)^d} \frac{1}{(-k^2)^2} . \tag{2.69}$$

We have to introduce some infrared regularization, such as a sharp cutoff or a non-zero mass; then the ultraviolet divergence of this integral is given by (1.29), and we obtain the $1/\varepsilon$ part of Λ^α:

$$\Lambda^\alpha = a(\mu)\frac{\alpha(\mu)}{4\pi\varepsilon}\gamma^\alpha \tag{2.70}$$

(it does not depend on momenta). This calculation can be viewed as the simplest case of the infrared rearrangement [Vladimirov (1980)]. The

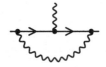

Fig. 2.9   One-loop QED vertex

renormalization constant

$$Z_\Gamma = 1 + a\frac{\alpha}{4\pi\varepsilon} \qquad (2.71)$$

is in agreement with (2.65), (2.53).

The bare charge $e_0 = Z_\alpha^{1/2}(\alpha(\mu))e(\mu)$ does not depend on μ. Differentiating (2.10) in $d\log\mu$, we obtain the RG equation for $\alpha(\mu)$:

$$\frac{d\log\alpha(\mu)}{d\log\mu} = -2\varepsilon - 2\beta(\alpha(\mu)), \qquad (2.72)$$

where β function is defined by

$$\beta(\alpha_s(\mu)) = \frac{1}{2}\frac{d\log Z_\alpha(\alpha_s(\mu))}{d\log\mu}. \qquad (2.73)$$

With one-loop accuracy, differentiating

$$Z_\alpha(\alpha) = 1 + z_1\frac{\alpha}{4\pi\varepsilon} + \cdots$$

we can retain only the leading term -2ε in (2.72):

$$\beta(\alpha) = \beta_0\frac{\alpha}{4\pi} + \cdots = -z_1\frac{\alpha}{4\pi} + \cdots$$

In other words, at one loop

$$Z_\alpha(\alpha) = 1 - \beta_0\frac{\alpha}{4\pi\varepsilon} + \cdots \qquad (2.74)$$

We have obtained (see (2.66), (2.32))

$$Z_\alpha = Z_A^{-1} = 1 + \frac{4}{3}\frac{\alpha}{4\pi\varepsilon} + \cdots \qquad (2.75)$$

Therefore, the QED β function is

$$\beta(\alpha) = \beta_0\frac{\alpha}{4\pi} + \cdots = -\frac{4}{3}\frac{\alpha}{4\pi} + \cdots \qquad (2.76)$$

When we consider the renormalized $\alpha(\mu)$ after taking the physical limit $\varepsilon \to 0$, we may omit -2ε in (2.72):

$$\frac{d\log\alpha(\mu)}{d\log\mu} = -2\beta(\alpha(\mu)). \qquad (2.77)$$

In QED $\beta_0 = -4/3$, and the β-function (2.76) is negative, at least at sufficiently small α, where perturbation theory is valid. Therefore, the running $\alpha(\mu)$ grows with μ. This corresponds to the physical picture of charge screening: at larger distances (smaller μ) the charge becomes smaller.

The RG equation (2.77) in the one-loop approximation,

$$\frac{d}{d\log\mu}\frac{\alpha(\mu)}{4\pi} = -2\beta_0\left(\frac{\alpha(\mu)}{4\pi}\right)^2,\qquad(2.78)$$

can be rewritten as

$$\frac{d}{d\log\mu}\frac{4\pi}{\alpha(\mu)} = 2\beta_0,\qquad(2.79)$$

and its solution is

$$\frac{4\pi}{\alpha(\mu')} - \frac{4\pi}{\alpha(\mu)} = 2\beta_0\log\frac{\mu'}{\mu}.$$

Finally,

$$\alpha(\mu') = \frac{\alpha(\mu)}{1 + 2\beta_0\dfrac{\alpha(\mu)}{4\pi}\log\dfrac{\mu'}{\mu}}.\qquad(2.80)$$

2.8 Electron mass

Until now we treated electrons as massless. This is a good approximation if characteristic energies are large (in the Z region at LEP, say). Now let's recall that they have mass. The QED Lagrangian (2.1) now is

$$L = \bar{\psi}_0\left(i\slashed{D} - m_0\right)\psi_0 + \cdots\qquad(2.81)$$

and the free electron propagator (2.5) becomes

$$S_0(p) = \frac{1}{\slashed{p} - m_0} = \frac{\slashed{p} + m_0}{p^2 - m_0^2}.\qquad(2.82)$$

In addition to (2.7), we have also mass renormalization

$$m_0 = Z_m(\alpha(\mu))\,m(\mu).\qquad(2.83)$$

The electron self-energy has two γ-matrix structures, because helicity is no longer conserved:

$$\Sigma(p) = \slashed{p}\Sigma_V(p^2) + m_0\Sigma_S(p^2),\qquad(2.84)$$

and the electron propagator is

$$S(p) = \frac{1}{\not{p} - m_0 - \not{p}\Sigma_V(p^2) - m_0\Sigma_S(p^2)} = \frac{1}{1 - \Sigma_V(p^2)} \frac{1}{\not{p} - \dfrac{1 + \Sigma_S(p^2)}{1 - \Sigma_V(p^2)}m_0}.$$
(2.85)

It should be equal to $Z_\psi S_r(p; \mu)$. The renormalization constants are determined by the conditions

$$(1 - \Sigma_V)Z_\psi = \text{finite}, \qquad \frac{1 + \Sigma_S}{1 - \Sigma_V}Z_m = \text{finite}.$$
(2.86)

The first of them is the same as in the massless case; the second one, defining Z_m, can be rewritten as

$$(1 + \Sigma_S)Z_\psi Z_m = \text{finite}.$$
(2.87)

The one-loop electron self-energy (Fig. 2.6) is

$$-i\Sigma(p) = \int \frac{d^d k}{(2\pi)^d} ie_0\gamma^\mu i\frac{\not{k} + \not{p} + m_0}{(k+p)^2 - m_0^2} ie_0\gamma^\nu \frac{-i}{k^2}\left(g_{\mu\nu} - \xi\frac{k_\mu k_\nu}{k^2}\right).$$
(2.88)

In order to single out Σ_S, we should retain the m_0 term in the numerator of the electron propagator (it flips helicity). At large p^2,

$$\begin{aligned}\Sigma_S(p^2) &= -ie_0^2 \int \frac{d^d k}{(2\pi)^d} \frac{\gamma^\mu\gamma^\nu}{(k+p)^2 k^2}\left(g_{\mu\nu} - \xi\frac{k_\mu k_\nu}{k^2}\right) \\ &= -ie_0^2(d - \xi) \int \frac{d^d k}{(2\pi)^d} \frac{1}{(k+p)^2 k^2}.\end{aligned}$$
(2.89)

Retaining only the ultraviolet divergence (1.29) and re-expressing via renormalized quantities (this is trivial at this order), we obtain

$$\Sigma_S = (3 + a(\mu))\frac{\alpha(\mu)}{4\pi\varepsilon}.$$
(2.90)

From (2.87) we see that a terms cancel:

$$(1 + \Sigma_S)Z_\psi Z_m = \left(1 + (3 + a)\frac{\alpha}{4\pi\varepsilon}\right)\left(1 - a\frac{\alpha}{4\pi\varepsilon}\right)Z_m = 1,$$

and Z_m is gauge-independent:

$$Z_m = 1 - 3\frac{\alpha}{4\pi\varepsilon} + \cdots$$
(2.91)

The bare mass $m_0 = Z_m(\alpha(\mu))m(\mu)$ does not depend on μ:

$$\frac{dm(\mu)}{d\log\mu} + \gamma_m(\alpha(\mu))m(\mu) = 0\,, \qquad (2.92)$$

where the mass anomalous dimension is

$$\gamma_m(\alpha(\mu)) = \frac{d\log Z_m(\alpha(\mu))}{d\log\mu}\,. \qquad (2.93)$$

From (2.91) we obtain

$$\gamma_m(\alpha) = \gamma_{m0}\frac{\alpha}{4\pi} + \cdots = 6\frac{\alpha}{4\pi} + \cdots \qquad (2.94)$$

In order to solve the RG equation (2.92) for $m(\mu)$, let's write it down together with the RG equation (2.72) for $\alpha(\mu)$:

$$\frac{d\log\alpha}{d\log\mu} = -2\beta(\alpha)\,, \qquad \frac{d\log m}{d\log\mu} = -\gamma_m(\alpha)\,,$$

and divide the second equation by the first one:

$$\frac{d\log m}{d\log\alpha} = \frac{\gamma_m(\alpha)}{2\beta(\alpha)}\,. \qquad (2.95)$$

The solution is

$$m(\mu') = m(\mu)\exp\int_{\alpha(\mu)}^{\alpha(\mu')}\frac{\gamma_m(\alpha)}{2\beta(\alpha)}\frac{d\alpha}{\alpha}\,. \qquad (2.96)$$

At one loop, we obtain from (2.76) and (2.94)

$$m(\mu') = m(\mu)\left(\frac{\alpha(\mu')}{\alpha(\mu)}\right)^{\gamma_{m0}/(2\beta_0)} = m(\mu)\left(\frac{\alpha(\mu')}{\alpha(\mu)}\right)^{-9/4}\,, \qquad (2.97)$$

The running electron mass $m(\mu)$ decreases with μ.

Chapter 3

QCD at one loop

3.1 Lagrangian and Feynman rules

The Lagrangian of QCD with n_f massless flavours is

$$L = \sum_i \bar{q}_{0i} i \slashed{D} q_{0i} - \frac{1}{4} G^a_{0\mu\nu} G^{a\mu\nu}_0 \, , \tag{3.1}$$

where q_{0i} are the quark fields,

$$D_\mu q_0 = (\partial_\mu - i g_0 A_{0\mu}) q_0 \, , \quad A_{0\mu} = A^a_{0\mu} t^a \tag{3.2}$$

are their covariant derivatives, $A^a_{0\mu}$ is the gluon field, t^a are the generators of the colour group, and the field strength tensor is defined by

$$[D_\mu, D_\nu] q_0 = -i g_0 G_{0\mu\nu} q_0 \, , \quad G_{0\mu\nu} = G^a_{0\mu\nu} t^a \, . \tag{3.3}$$

It is given by

$$G^a_{0\mu\nu} = \partial_\mu A^a_{0\nu} - \partial_\nu A^a_{0\mu} + g_0 f^{abc} A^b_{0\mu} A^c_{0\nu} \, , \tag{3.4}$$

where the structure constants f^{abc} are defined by

$$[t^a, t^b] = i f^{abc} t^c \, . \tag{3.5}$$

Due to the gauge invariance, it is not possible to obtain the gluon propagator from this Lagrangian. We should fix the gauge. The most popular way (called the covariant gauge) requires to add the gauge-fixing term (with the parameter a_0) and the ghost term to the Lagrangian (3.1):

$$\Delta L = -\frac{1}{2a_0} \left(\partial_\mu A^{a\mu}_0 \right)^2 + (\partial^\mu \bar{c}^a_0)(D_\mu c^a_0) \, . \tag{3.6}$$

43

Here c_0^a is the ghost field — a scalar field obeying Fermi statistics, with the colour index a (like the gluon). Its covariant derivative is

$$D_\mu c_0^a = \left(\partial_\mu \delta^{ab} - ig_0 A_{0\mu}^{ab} \right) c_0^b, \quad A_{0\mu}^{ab} = A_{0\mu}^c (t^c)^{ab}, \tag{3.7}$$

where

$$(t^c)^{ab} = i f^{acb} \tag{3.8}$$

are the generators of the colour group in the adjoint representation.

$$
\begin{aligned}
&\begin{array}{c}\bullet\!\!\xrightarrow{\hspace{1.5cm}}\!\!\bullet \\ p\end{array} \qquad = i S_0(p) \\[8pt]
&\begin{array}{c}\overset{a}{\underset{\mu}{\bullet}}\!\!\!\wwww\!\!\!\overset{b}{\underset{\nu}{\bullet}} \\ p\end{array} \qquad = -i\delta^{ab} D_{\mu\nu}^0(p) \\[8pt]
&\begin{array}{c}\overset{a}{\bullet}\!\!-\!\!-\!\!\to\!\!-\!\!-\!\!\overset{b}{\bullet} \\ p\end{array} \qquad = i\delta^{ab} G_0(p)
\end{aligned}
$$

Fig. 3.1   Propagators in QCD

The quadratic part of the QCD Lagrangian gives the propagators shown in Fig. 3.1. The quark propagator is $S_0(p)$ (2.5); the unit colour matrix is assumed. The gluon propagator is $D_{\mu\nu}^0(p)$ (2.6); the unit matrix in the adjoint representation δ^{ab} is written down explicitly. The ghost propagator is

$$G_0(p) = \frac{1}{p^2} \tag{3.9}$$

times δ^{ab}.

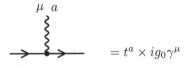

Fig. 3.2   Quark–gluon vertex

The quark–gluon vertex is shown in Fig. 3.2. The problem of calculation of a QCD Feynman diagram reduces to two separate subproblems:

- calculation of the colour factor;

• calculation of the "colourless" diagram.

The first subproblem can be formulated as calculation of the *colour diagram*. It looks like the original QCD diagram, where quark lines and gluon lines mean the unit matrices in the fundamental representation and in the adjoint one; ghost lines can be replaced by gluon lines in colour diagrams. The quark–gluon vertex in a colour diagram gives t^a, the first factor in Fig. 3.2. When calculating the "colourless" diagram, all factors except the colour structure are included; the quark–gluon vertex gives $ig_0\gamma^\mu$, the second factor in Fig. 3.2.

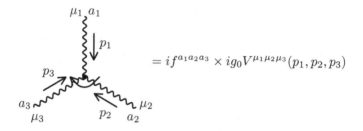

$$= if^{a_1 a_2 a_3} \times ig_0 V^{\mu_1\mu_2\mu_3}(p_1, p_2, p_3)$$

Fig. 3.3   Three-gluon vertex

The three-gluon vertex is shown in Fig. 3.3. When separating it into the colour structure (which appears in the colour diagram) and the Lorentz structure (which appears in the "colourless" diagram), we have to choose some rotation direction. These rotation directions must be the same in the colour diagram and in the "colourless" one. The colour structure is $if^{a_1 a_2 a_3}$, where the colour indices are written in the chosen order (in Fig. 3.3, clockwise). The Lorentz structure is

$$V^{\mu_1\mu_2\mu_3}(p_1, p_2, p_3) = (p_3 - p_2)^{\mu_1} g^{\mu_2\mu_3} + (p_1 - p_3)^{\mu_2} g^{\mu_3\mu_1} + (p_2 - p_1)^{\mu_3} g^{\mu_1\mu_2},$$

$$(3.10)$$

where the polarization indices and the momenta are numbered in the same way, here clockwise. Of course, the full three-gluon vertex does not know about the rotation direction: if we reverse it, both the colour factor and the V tensor change sign, and their product does not change.

The four-gluon vertex contains terms with three different colour structures. Therefore, any diagram containing at least one four-gluon vertex cannot be factorized into a colour diagram and a "colourless" one. This is very inconvenient for programs automating diagram calculations. The

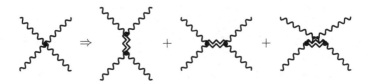

Fig. 3.4 Four-gluon vertex

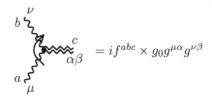
$$= i\delta^{ab}(g_{\mu\alpha}g_{\nu\beta} - g_{\mu\beta}g_{\nu\alpha})$$

Fig. 3.5 Propagator of the auxiliary field

$$= if^{abc} \times g_0 g^{\mu\alpha} g^{\nu\beta}$$

Fig. 3.6 Vertex of the auxiliary field interaction with gluons

authors of (at least) two such programs (CompHEP and GEFICOM) invented the same trick to ensure the possibility of calculating the colour factor as a separate subproblem. Let's say that there is no four-gluon vertex in QCD, but there is a new field interacting with gluons (this field is shown as a double zigzag line in Fig. 3.4). This is an antisymmetric tensor field; its propagator is shown in Fig. 3.5, and the vertex of its interaction with gluons — in Fig. 3.6. Its propagator in the momentum space does not depend on p. Therefore, in the coordinate space, it contains $\delta(x)$ — this particle does not propagate, and two vertices of its interaction with gluons (Fig. 3.4) are at the same point. In the colour diagram, lines of this particle can be replaces by gluon lines, and vertices — by three-gluon vertices. In "colourless" diagrams, the second factor in Fig. 3.6 is used for the vertices. The rotation directions must agree, as usual (if the rotation direction is reversed, the two polarization indices of the tensor field are interchanges, and this gives a minus sign). For example, one of the three contributions

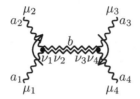

Fig. 3.7 One of the contributions to the four-gluon vertex

to the four-gluon vertex (shown in Fig. 3.7) is

$$i f^{a_1 a_2 b} i f^{a_3 a_4 b} \times g_0 g^{\mu_1 \nu_1} g^{\mu_2 \nu_2} i \left(g_{\nu_1 \nu_3} g_{\nu_2 \nu_4} - g_{\nu_1 \nu_4} g_{\nu_2 \nu_3} \right) g_0 g^{\mu_3 \nu_3} g^{\mu_4 \nu_4}$$
$$= i f^{a_1 a_2 b} i f^{a_3 a_4 b} \times i g_0^2 \left(g^{\mu_1 \mu_3} g^{\mu_2 \mu_4} - g^{\mu_1 \mu_4} g^{\mu_2 \mu_3} \right) .$$

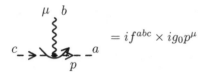

Fig. 3.8 Ghost–gluon vertex

Finally, the ghost–gluon vertex is shown in Fig. 3.8. It contains only the outgoing ghost momentum, but not the incoming ghost momentum. This is because its Lagrangian (3.6) contains the covariant derivative of c but the ordinary derivative of $\bar{c}$. In colour diagrams, we replace ghost lines by gluon lines, and vertices — by three-gluon vertices; their rotation direction is fixed: the incoming ghost $\to$ the outgoing ghost $\to$ the gluon (Fig. 3.8).

Calculation of colour diagrams is discussed in Appendix A.

The renormalized fields and parameters are related to the bare ones, similarly to QED (2.7), by

$$q_{i0} = Z_q^{1/2} q_i, \quad A_0 = Z_A^{1/2} A, \quad c_0 = Z_c^{1/2} c, \quad a_0 = Z_A a, \quad g_0 = Z_\alpha^{1/2} g;$$
(3.11)

the QCD running coupling $\alpha_s(\mu)$ is

$$\frac{\alpha_s(\mu)}{4\pi} = \mu^{-2\varepsilon} \frac{g^2}{(4\pi)^{d/2}} e^{-\gamma\varepsilon}, \quad \frac{g_0^2}{(4\pi)^{d/2}} = \mu^{2\varepsilon} \frac{\alpha_s(\mu)}{4\pi} Z_\alpha(\alpha(\mu)) e^{\gamma\varepsilon}. \quad (3.12)$$

3.2 Quark propagator

The quark propagator in QCD has the same structure (2.39) as the electron propagator in QED (Sect. 2.6). In massless QCD, the quark self-energy has the form

$$\Sigma(p) = \not{p}\Sigma_V(p^2).$$ (3.13)

The one-loop diagram (Fig. 3.9) differs from the QED one (2.50) only by the replacement $e_0 \to g_0$ and by the colour factor C_F (A.25):

$$\Sigma_V(p^2) = -C_F \frac{g_0^2(-p^2)^{-\varepsilon}}{(4\pi)^{d/2}} \frac{d-2}{2} a_0 G_1.$$ (3.14)

Therefore, the quark field renormalization constant with one-loop accuracy is

$$Z_q = 1 - C_F a \frac{\alpha_s}{4\pi\varepsilon} + \cdots$$ (3.15)

and the quark-field anomalous dimension is

$$\gamma_q = 2C_F a \frac{\alpha_s}{4\pi} + \cdots$$ (3.16)

Fig. 3.9   One-loop quark self-energy

If the quark flavour q_i has a non-zero mass, its self-energy has also the helicity-flipping structure $\Sigma_S(p^2)$. The calculation of its one-loop ultraviolet-divergent part (Sect. 2.8) applies in QCD practically unchanged, with the substitution $\alpha \to \alpha_s$ and the extra colour factor C_F:

$$\Sigma_S = C_F(3+a) \frac{\alpha_s}{4\pi\varepsilon} + \cdots$$ (3.17)

Therefore, the quark-mass renormalization constant is

$$Z_m = 1 - 3C_F \frac{\alpha_s}{4\pi\varepsilon} + \cdots$$ (3.18)

and the mass anomalous dimension is

$$\gamma_m = 6C_F \frac{\alpha_s}{4\pi} + \cdots$$ (3.19)

3.3 Gluon propagator

Ward identities in QCD are more complicated than in QED. Nevertheless, the gluon self-energy $i\delta^{ab}\Pi_{\mu\nu}(p)$ is transverse:

$$\Pi_{\mu\nu}(p)p^\mu = 0, \quad \Pi_{\mu\nu}(p) = (p^2 g_{\mu\nu} - p_\mu p_\nu)\Pi(p^2), \tag{3.20}$$

as in QED. Explanation of this is beyond the level of these lectures.

Fig. 3.10   One-loop gluon self-energy

At one loop, the gluon self-energy is given by three diagrams (Fig. 3.10). The quark-loop contribution differs from the QED result (2.27) only by the substitution $e_0 \to g_0$, and the colour factor $T_F n_f$:

$$\Pi_q(p^2) = -T_F n_f \frac{g_0^2 (-p^2)^{-\varepsilon}}{(4\pi)^{d/2}} 2\frac{d-2}{d-1} G_1. \tag{3.21}$$

The gluon-loop contribution (Fig. 3.11) has the symmetry factor $\frac{1}{2}$ and the colour factor C_A (A.31). In the Feynman gauge $a_0 = 1$,

$$\Pi_1{}_\mu^\mu = -i\frac{1}{2}C_A g_0^2 \int \frac{d^d k}{(2\pi)^d} \frac{N}{k^2(k+p)^2}, \tag{3.22}$$
$$N = V_{\mu\alpha\beta}(p, -k-p, k)V^{\mu\beta\alpha}(-p, -k, k+p).$$

The Lorentz part of the three-gluon vertex (3.10) here is

$$V_{\mu\alpha\beta}(p, -k-p, k) = (2k+p)_\mu g_{\alpha\beta} - (k-p)_\alpha g_{\beta\mu} - (k+2p)_\beta g_{\mu\alpha}.$$

$V^{\mu\beta\alpha}(-p, -k, k+p)$ coincides with it. Therefore, the numerator in (3.22) is

$$N = d\left[(2k+p)^2 + (k-p)^2 + (k+2p)^2\right]$$
$$- 2(2k+p)\cdot(k-p) - 2(2k+p)\cdot(k+2p) + 2(k+2p)\cdot(k-p).$$

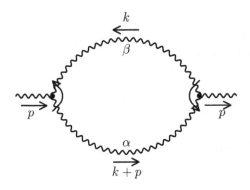

Fig. 3.11   Gluon-loop contribution

Using the "multiplication table" (2.25) and omitting terms with $D_{1,2}$ in the numerator (which give vanishing integrals), we have

$$p^2 = -1, \quad k^2 = -D_2 \Rightarrow 0, \quad p \cdot k = \frac{1}{2}(1 - D_1 + D_2) \Rightarrow \frac{1}{2},$$

$$N \Rightarrow -3(d-1).$$

Finally, we arrive at

$$\Pi_{1\mu}^{\mu} = -\frac{3}{2}C_A \frac{g_0^2(-p^2)^{1-\varepsilon}}{(4\pi)^{d/2}} G_1(d-1). \tag{3.23}$$

The ghost-loop contribution (Fig. 3.12) has the fermion-loop factor -1 and the colour factor C_A (A.31):

$$\Pi_{2\mu}^{\mu} = iC_A g_0^2 \int \frac{d^d k}{(2\pi)^d} \frac{k \cdot (k+p)}{k^2(k+p)^2} = -\frac{1}{2}C_A \frac{g_0^2(-p^2)^{1-\varepsilon}}{(4\pi)^{d/2}} G_1. \tag{3.24}$$

These two contributions, taken separately, are not transverse. Their sum is transverse, and hence has the structure (3.20):

$$\Pi_g(p^2) = -\frac{\Pi_{1\mu}^{\mu} + \Pi_{2\mu}^{\mu}}{(d-1)(-p^2)} = C_A \frac{g_0^2(-p^2)^{-\varepsilon}}{(4\pi)^{d/2}} G_1 \frac{3d-2}{2(d-1)}. \tag{3.25}$$

In an arbitrary covariant gauge,

$$\Pi_g(p^2) = C_A \frac{g_0^2(-p^2)^{-\varepsilon}}{(4\pi)^{d/2}} \frac{G_1}{2(d-1)}$$
$$\times \left[3d - 2 + (d-1)(2d-7)\xi - \frac{1}{4}(d-1)(d-4)\xi^2 \right] \tag{3.26}$$

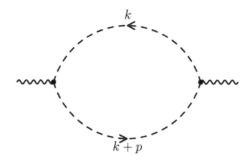

Fig. 3.12   Ghost-loop contribution

(here $\xi = 1 - a_0$ (2.45); we leave the derivation as an exercise for the reader).

The transverse part of the gluon propagator with one-loop accuracy, expressed via renormalized quantities and expanded in ε, is

$$p^2 D_\perp(p^2) = 1 + \frac{\alpha_s(\mu)}{4\pi\varepsilon} e^{-L\varepsilon} \left[-\frac{1}{2}\left(a - \frac{13}{3} \right) C_A - \frac{4}{3} T_F n_f \right.$$
$$\left. + \left(\frac{9a^2 + 18a + 97}{36} C_A - \frac{20}{9} T_F n_f \right) \varepsilon \right], \tag{3.27}$$

where $L = \log(-p^2)/\mu^2$. Therefore, the gluon field renormalization constant is

$$Z_A = 1 - \frac{\alpha_s}{4\pi\varepsilon} \left[\frac{1}{2}\left(a - \frac{13}{3} \right) C_A + \frac{4}{3} T_F n_f \right] + \cdots \tag{3.28}$$

We don't write down the expression for the renormalized gluon propagator; it is easy to do this. The gluon field anomalous dimension is

$$\gamma_A = \left[\left(a - \frac{13}{3} \right) C_A + \frac{8}{3} T_F n_f \right] \frac{\alpha_s}{4\pi} + \cdots \tag{3.29}$$

Making the substitutions $\alpha_s \to \alpha$, $C_A \to 0$, $T_F n_f \to 1$, we reproduce the QED results (2.32), (2.36).

3.4   Ghost propagator

The ghost propagator has the structure

$$iG(p) = iG_0(p) + iG_0(p)(-i)\Sigma iG_0(p) + iG_0(p)(-i)\Sigma iG_0(p)(-i)\Sigma iG_0(p) + \cdots$$
$$(3.30)$$

where the ghost self-energy $-i\Sigma$ is a scalar function of p^2, and the free
ghost propagator $G_0(p)$ is given by (3.9). Therefore,

$$G(p) = \frac{1}{p^2 - \Sigma(p^2)} \,. \tag{3.31}$$

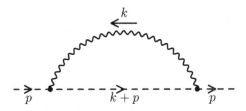

Fig. 3.13   One-loop ghost self-energy

At one loop (Fig. 3.13),

$$\begin{aligned}
\Sigma(p^2) &= -iC_A g_0^2 \int \frac{d^d k}{(2\pi)^d} \frac{p^\mu (k+p)^\nu}{k^2 (k+p)^2} \left(g_{\mu\nu} - \xi \frac{k_\mu k_\nu}{k^2} \right) \\
&= C_A \frac{g_0^2 (-p^2)^{1-\varepsilon}}{(4\pi)^{d/2}} \left[-\frac{1}{2} G(1,1) + \frac{\xi}{4} G(1,2) \right] \\
&= -\frac{1}{4} C_A \frac{g_0^2 (-p^2)^{1-\varepsilon}}{(4\pi)^{d/2}} G_1 \left[d - 1 - (d-3)a_0 \right] .
\end{aligned} \tag{3.32}$$

The propagator, expressed via renormalized quantities and expanded in ε,
is

$$G(p) = \frac{1}{p^2} \left[1 + C_A \frac{\alpha_s(\mu)}{4\pi\varepsilon} e^{-L\varepsilon} \frac{3 - a + 4\varepsilon}{4} \right] . \tag{3.33}$$

Therefore, the ghost field renormalization constant is

$$Z_c = 1 + C_A \frac{3 - a}{4} \frac{\alpha_s}{4\pi\varepsilon} + \cdots \tag{3.34}$$

and the anomalous dimension is

$$\gamma_c = -C_A \frac{3-a}{2} \frac{\alpha_s}{4\pi} + \cdots \qquad (3.35)$$

3.5 Quark–gluon vertex

Let's find the ultraviolet divergence of the quark–gluon vertex at one loop (Fig. 3.14). The first diagram differs from the QED one only by the colour factor $C_F - C_A/2$ (A.34), and

$$\Lambda_1^\alpha = \left(C_F - \frac{C_A}{2} \right) \frac{\alpha_s}{4\pi\varepsilon} \gamma^\alpha . \qquad (3.36)$$

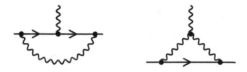

Fig. 3.14   Quark–gluon vertex at one loop

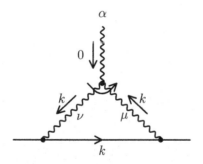

Fig. 3.15   The non-abelian diagram

The second diagram (Fig. 3.15) has the colour factor $C_A/2$ (A.36). It has a logarithmic ultraviolet divergence. Therefore, as in Sect. 2.7, in order to find the $1/\varepsilon$ term we may nullify all the external momenta and provide some infrared cutoff. In the Feynman gauge ($a_0 = 1$),

$$\Lambda_2^\alpha = i \frac{C_A}{2} g_0^2 \int \frac{d^d k}{(2\pi)^d} \frac{\gamma_\mu \slashed{k} \gamma_\nu}{(k^2)^3} V^{\alpha\nu\mu}(0, -k, k), \tag{3.37}$$

where the Lorentz part of the three-gluon vertex (3.10) is

$$V^{\alpha\nu\mu}(0, -k, k) = 2k^\alpha g^{\mu\nu} - k^\mu g^{\nu\alpha} - k^\nu g^{\mu\alpha}.$$

Therefore,

$$\Lambda_2^\alpha = i \frac{C_A}{2} g_0^2 \int \frac{d^d k}{(2\pi)^d} \frac{2\gamma_\mu \slashed{k} \gamma^\mu k^\alpha - 2k^2 \gamma^\alpha}{(k^2)^3}.$$

Averaging over k directions:

$$\slashed{k} k^\alpha \to \frac{k^2}{d} \gamma^\alpha$$

and using the ultraviolet divergence (1.29), we obtain

$$\Lambda_2^\alpha = \frac{3}{2} C_A \frac{\alpha_s}{4\pi\varepsilon} \gamma^\alpha. \tag{3.38}$$

In the arbitrary covariant gauge

$$\Lambda_2^\alpha = \frac{3}{4}(1 + a) C_A \frac{\alpha_s}{4\pi\varepsilon} \gamma^\alpha \tag{3.39}$$

(derive this result!).

The $1/\varepsilon$ term (the ultraviolet divergence) of the one-loop quark–gluon vertex is, from (3.36) and (3.39),

$$\Lambda^\alpha = \left(C_F a + C_A \frac{a+3}{4}\right) \frac{\alpha_s}{4\pi\varepsilon} \gamma^\alpha. \tag{3.40}$$

Therefore, the quark–gluon vertex renormalization constant is

$$Z_\Gamma = 1 + \left(C_F a + C_A \frac{a+3}{4}\right) \frac{\alpha_s}{4\pi\varepsilon} + \cdots \tag{3.41}$$

3.6 Coupling constant renormalization

The coupling renormalization constant is

$$Z_\alpha = (Z_\Gamma Z_q)^{-2} Z_A^{-1}, \tag{3.42}$$

similarly to Sect. 2.7. The product $Z_\Gamma Z_q$, with one-loop accuracy, is, from (3.41) and (3.15),

$$Z_\Gamma Z_q = 1 + C_A \frac{a+3}{4} \frac{\alpha_s}{4\pi\varepsilon} + \cdots \tag{3.43}$$

In QED, it was equal to 1 (2.65), due to the Ward identity (Sect. 2.7). Ward identities in QCD are more complicated. Making the replacements $\alpha_s \to \alpha$, $C_F \to 1$, $C_A \to 0$, we reproduce the simple QED result.

The coupling renormalization constant, with one-loop accuracy, is, from (3.43) and (3.28),

$$Z_\alpha = 1 - \left(\frac{11}{3} C_A - \frac{4}{3} T_F n_f \right) \frac{\alpha_s}{4\pi\varepsilon} + \cdots \tag{3.44}$$

It is gauge-invariant. This is a strong check of our calculations[1]. The β-function (2.73) in QCD is, with one-loop accuracy,

$$\beta(\alpha_s) = \beta_0 \frac{\alpha_s}{4\pi} + \cdots \qquad \beta_0 = \frac{11}{3} C_A - \frac{4}{3} T_F n_f . \tag{3.45}$$

If there are not too many quark flavours (namely, for $N_c = 3$, if $n_f \leq 16$), then $\beta_0 > 0$, in contrast to the QED case $\beta_0 = -4/3$.

The RG equation

$$\frac{d \log \alpha_s(\mu)}{d \log \mu} = -2\beta(\alpha_s(\mu)) \tag{3.46}$$

shows that $\alpha_s(\mu)$ decreases when μ increases. This behaviour (opposite to screening) is called *asymptotic freedom*. Keeping only the leading (one-loop) term in the β-function, we can rewrite the RG equation in the form (2.79). The running coupling at the renormalization scale μ' is related to that at μ by (2.80), but now with positive β_0. The solution of the equation (2.79) can also be written as

$$\alpha_s(\mu) = \frac{2\pi}{\beta_0 \log \dfrac{\mu}{\Lambda_{\overline{\mathrm{MS}}}}} , \tag{3.47}$$

where $\Lambda_{\overline{\mathrm{MS}}}$ plays the role of the integration constant. It sets the characteristic energy scale of QCD; the coupling becomes small at $\mu \gg \Lambda_{\overline{\mathrm{MS}}}$.

Coupling-constant renormalization in the non-abelian gauge theory was first considered by [Vanyashin and Terentev (1965)], and the sign of the

---

[1]We have actually presented detailed calculations of Z_A and Z_Γ only in the Feynman gauge $a_0 = 1$; this is enough for obtaining (3.44).

β-function corresponding to asymptotic freedom was obtained. The magnitude was not quite right (it is right for a spontaneously-broken gauge theory with a higgs). The first correct calculation of the one-loop β-function in the non-abelian gauge theory has been published by [Khriplovich (1969)]. It was done in the Coulomb gauge, which is ghost-free. The contribution of the transverse-gluon loop has the same sign as that of the quark loop, i.e., leads to screening. However, there is another contribution, that of the loop with an instantaneous Coulomb gluon. It has the opposite sign, and outweights the first contribution. This β-function has also been calculated by ['t Hooft (1971)], but not published (mentioned after Symanzik's talk at a meeting in Marseilles in 1972). Later it was calculated in the famous papers [Gross and Wilczek (1973); Politzer (1973)]. The authors of these papers applied asymptotic freedom to explain the observed behaviour of deep inelastic electron–proton scattering. This was the real beginning of QCD as a theory of strong interactions. They received the Nobel prize in 2004.

3.7   Ghost–gluon vertex

All QCD vertices contain just one coupling constant. Its renormalization can be found from renormalization of any vertex: quark–gluon (Sect. 3.5, 3.6), ghost–gluon, three-gluon, or four-gluon. Here we shall derive it again by calculating the ghost–gluon vertex at one loop.

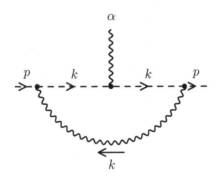

Fig. 3.16   One-loop ghost–gluon vertex (diagram 1)

The first diagram (Fig. 3.16) has the colour factor $C_A/2$ (A.38). The ultraviolet $1/\varepsilon$ divergence of this diagram is proportional to the bare vertex (i.e., to the outgoing ghost momentum p^α); the coefficient diverges logarithmically. We may nullify the external momenta except p, and use (1.29):

$$
\begin{aligned}
\Lambda_1^\alpha &= -i\frac{C_A}{2}g_0^2 \int \frac{d^dk}{(2\pi)^d}\frac{k^\alpha p^\mu k^\nu}{(k^2)^3}\left(g_{\mu\nu} - \xi\frac{k_\mu k_\nu}{k^2}\right) \\
&= -i\frac{C_A}{2}g_0^2 a_0 \int \frac{d^dk}{(2\pi)^d}\frac{k^\alpha p\cdot k}{(k^2)^3} \\
&= -i\frac{C_A}{2}g_0^2 a_0 \frac{1}{4}p^\alpha \int \frac{d^dk}{(2\pi)^d}\frac{1}{(k^2)^2} = \frac{1}{8}C_A a\frac{\alpha_s}{4\pi\varepsilon}p^\alpha .
\end{aligned}
\tag{3.48}
$$

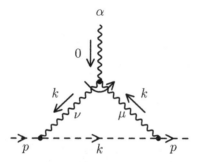

Fig. 3.17   One-loop ghost–gluon vertex (diagram 2)

The second diagram (Fig. 3.17) has the same colour factor; in the Feynman gauge ($a_0 = 1$),

$$
\begin{aligned}
\Lambda_2^\alpha &= i\frac{C_A}{2}g_0^2 \int \frac{d^dk}{(2\pi)^d}\frac{p_\mu k_\nu}{(k^2)^3}V^{\alpha\nu\mu}(0, -k, k) \\
&= i\frac{C_A}{2}g_0^2 \int \frac{d^dk}{(2\pi)^d}\frac{p\cdot k\, k^\alpha - k^2 p^\alpha}{(k^2)^3} \\
&= -\frac{3}{4}i\frac{C_A}{2}g_0^2 p^\alpha \int \frac{d^dk}{(2\pi)^d}\frac{1}{(k^2)^2} = \frac{3}{8}C_A\frac{\alpha_s}{4\pi\varepsilon}p^\alpha .
\end{aligned}
\tag{3.49}
$$

In the arbitrary covariant gauge,

$$
\Lambda_2^\alpha = \frac{3}{8}C_A a\frac{\alpha_s}{4\pi\varepsilon}p^\alpha
\tag{3.50}
$$

(derive this result!).

The full result for the $1/\varepsilon$ term in the ghost–gluon vertex at one loop is

$$
\Lambda^\alpha = \frac{1}{2}C_A a\frac{\alpha_s}{4\pi\varepsilon}p^\alpha .
\tag{3.51}
$$

Therefore, the vertex renormalization factor is

$$Z_{\Gamma c} = 1 + \frac{1}{2}C_A a \frac{\alpha_s}{4\pi\varepsilon} + \cdots \qquad (3.52)$$

The coupling renormalization constant

$$Z_\alpha = (Z_{\Gamma c} Z_c)^{-2} Z_A^{-1} \qquad (3.53)$$

is, from

$$Z_{\Gamma c} Z_c = 1 + C_A \frac{3+a}{4} \frac{\alpha_s}{4\pi\varepsilon}$$

(which follows from (3.34) and (3.52)), equal to (3.44). Thus we have re-derived β_0 (3.45); this derivation is, probably, slightly easier than that from the quark propagator and quark–gluon vertex.

3.8 Three-gluon vertex

The three-gluon vertex (the sum of one-particle-irreducible diagrams not including the external propagators, Fig. 3.18) has the structure[2]

$$\begin{aligned} ig_0 \Gamma^{a_1 a_2 a_3}_{\mu_1\mu_2\mu_3}(p_1, p_2, p_3) &= if^{a_1 a_2 a_3} \\ &\times ig_0 \left[V_{\mu_1\mu_2\mu_3}(p_1, p_2, p_3) + \Lambda_{\mu_1\mu_2\mu_3}(p_1, p_2, p_3) \right], \end{aligned} \qquad (3.54)$$

where the leading-order vertex is given by (3.10). Due to the Bose symmetry, $\Lambda_{\mu_1\mu_2\mu_3}(p_1, p_2, p_3)$ is antisymmetric with respect to interchanges of 1, 2, and 3. It can be written as a sum of tensor structures (constructed from

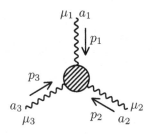

Fig. 3.18   Three-gluon vertex

---

[2]The colour structure $d^{a_1 a_2 a_3}$ can also appear at higher orders, but it does not appear at the one-loop level.

$g_{\mu\nu}$ and p_i^μ) multiplied by scalar functions of p_1^2, p_2^2, p_3^2, see, e.g., [Davy-dychev *et al.* (1996)]. In the one-loop approximation, $\Lambda_{\mu_1\mu_2\mu_3}(p_1, p_2, p_3)$ is given by the diagrams in Fig. 3.19. Here we shall calculate the ultraviolet-divergent part of the one-loop vertex in the Feynman gauge. The full one-loop result in an arbitrary covariant gauge has been obtained in [Davy-dychev *et al.* (1996)].

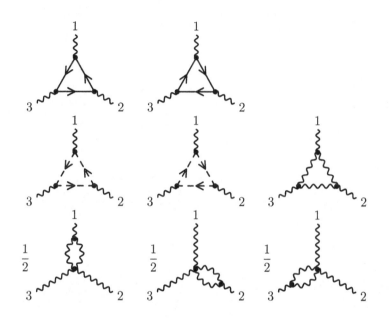

Fig. 3.19   One-loop diagrams for the three-gluon vertex

Let us consider the quark-loop contribution (Fig. 3.20) first. The "colourless" factor of the second diagram, which has the opposite direction of the quark line, has an extra minus sign as compared to the first diagram, because each of three quark propagators changes its sign. Therefore, the colour structure of the quark contribution is $if^{a_1 a_2 a_3} T_F$, see (A.28). We should also multiply it by the number of flavours n_f.

It is easy to see that the choice of momenta in Fig. 3.20 is consistent, due to $p_1 + p_2 + p_3 = 0$. The contribution of the diagram in Fig. 3.20 plus the diagram with the opposite direction of the quark line to the three-gluon

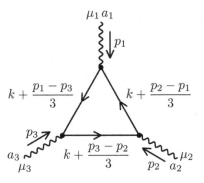

Fig. 3.20   The quark-loop contribution to the three-gluon vertex

vertex (3.54) is

$$\Lambda_q^{\mu_1\mu_2\mu_3}(p_1,p_2,p_3) = -iT_F n_f g_0^2 \int \frac{d^d k}{(2\pi)^d}$$

$$\times \frac{\text{Tr}\, \gamma^{\mu_1}\left(\slashed{k} + \dfrac{\slashed{p}_2 - \slashed{p}_1}{3}\right)\gamma^{\mu_2}\left(\slashed{k} + \dfrac{\slashed{p}_3 - \slashed{p}_2}{3}\right)\gamma^{\mu_3}\left(\slashed{k} + \dfrac{\slashed{p}_1 - \slashed{p}_3}{3}\right)}{\left(k + \dfrac{p_2 - p_1}{3}\right)^2 \left(k + \dfrac{p_3 - p_2}{3}\right)^2 \left(k + \dfrac{p_1 - p_3}{3}\right)^2} . \qquad (3.55)$$

We are interested only in the ultraviolet-divergent part of this contribution. It seems that the most divergent part is the one with three factors
k in the numerator (its ultraviolet divergence is linear). This part contains
the tensor

$$T^{\alpha_1\alpha_2\alpha_3}(p_1,p_2,p_3)$$

$$= \int \frac{k^{\alpha_1} k^{\alpha_2} k^{\alpha_3}\, d^d k}{\left(k + \dfrac{p_2 - p_1}{3}\right)^2 \left(k + \dfrac{p_3 - p_2}{3}\right)^2 \left(k + \dfrac{p_1 - p_3}{3}\right)^2} . \qquad (3.56)$$

This tensor is symmetric in its indices α_1, α_2, α_3; it is antisymmetric with
respect to interchanges of p_1, p_2, p_3 (this can be easily checked using the
substitution $k \to -k$). It can be written as a sum of tensor structures
multiplied by scalar functions of p_1^2, p_2^2, p_3^2. One of such terms is

$$f(p_1^2,p_2^2,p_3^2)\,(p_1^{\alpha_1} g^{\alpha_2\alpha_3} + p_1^{\alpha_2} g^{\alpha_3\alpha_1} + p_1^{\alpha_3} g^{\alpha_1\alpha_2}) + (\text{cycle}),$$

where $f(p_1^2,p_2^2,p_3^2) = -f(p_1^2,p_3^2,p_2^2)$, and (cycle) means cyclic permutations of 1, 2, 3. The function f has dimensionality -2ε; however, due

to its antisymmetry, it cannot contain a term $\mathrm{const}/\varepsilon$ in the limit $\varepsilon \to 0$. All the other tensor structures contain 3 vectors p_i, and their coefficients have dimensionality $-2 - 2\varepsilon$. Therefore, these coefficients are given by ultraviolet-convergent integrals. We conclude that the tensor integral (3.56) is ultraviolet-convergent.

This argument can be re-formulated in a slightly different form. The scalar coefficients in the expansion of the tensor integral (3.56) in tensor structures are obtained by solving a linear system from (scalar) contractions of $T^{\alpha_1\alpha_2\alpha_3}$ with terms constructed from $g_{\mu\nu}$ and $p_{i\mu}$. It is not difficult to check that $T^{\alpha_1\alpha_2\alpha_3} p_{3\alpha_3}$ is ultraviolet-convergent (and hence contractions with p_2 and p_3 are finite too). Similarly, the contraction with $g_{\alpha_1\alpha_2}$ is convergent.

Therefore, the only ultraviolet-divergent terms in (3.55) are those with two factors k in the numerator. We are only interested in the $1/\varepsilon$ terms; therefore, we may nullify all external momenta (of course, we'll have to introduce some infrared regularization, otherwise the integral vanishes); we may do Dirac algebra in four dimensions, and average over k directions: $k^\mu k^\nu \to (k^2/4)g^{\mu\nu}$. Then the trace in the numerator becomes

$$\frac{4}{3}k^2 V^{\mu_1\mu_2\mu_3}(p_1, p_2, p_3)\,,$$

and using (1.29) we arrive at

$$\Lambda_q^{\mu_1\mu_2\mu_3}(p_1, p_2, p_3) - \left[\frac{4}{3}T_F n_f V^{\mu_1\mu_2\mu_3}(p_1, p_2, p_3) + \mathcal{O}(\varepsilon)\right]\frac{\alpha_s}{4\pi\varepsilon}\,. \tag{3.57}$$

Now we turn to the ghost-loop contribution (Fig. 3.21). The colour factor of this diagram is $C_A/2$ (A.38). Adding also the contribution of the diagram with the opposite direction of the ghost line (whose colour factor is $-C_A/2$), we have

$$\Lambda_c^{\mu_1\mu_2\mu_3}(p_1, p_2, p_3) = -i\frac{C_A}{2}g_0^2$$
$$\times \int \frac{d^d k}{(2\pi)^d} \frac{1}{\left(k + \dfrac{p_2 - p_1}{3}\right)^2 \left(k + \dfrac{p_3 - p_2}{3}\right)^2 \left(k + \dfrac{p_1 - p_3}{3}\right)^2}$$

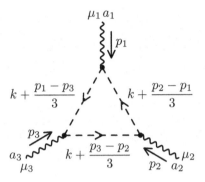

Fig. 3.21   The ghost-loop contribution to the three-gluon vertex

$$\times \left[\left(k + \frac{p_1 - p_3}{3} \right)^{\mu_1} \left(k + \frac{p_2 - p_1}{3} \right)^{\mu_2} \left(k + \frac{p_3 - p_2}{3} \right)^{\mu_3} \right.$$
$$\left. + \left(k + \frac{p_2 - p_1}{3} \right)^{\mu_1} \left(k + \frac{p_3 - p_2}{3} \right)^{\mu_2} \left(k + \frac{p_1 - p_3}{3} \right)^{\mu_3} \right]. \qquad (3.58)$$

Terms with three factors k in the numerator (3.56) don't contribute. In terms with two factors k, we average over k directions, and the numerator becomes $-V^{\mu_1 \mu_2 \mu_3}(p_1, p_2, p_3)/12$. The result is

$$\Lambda_c^{\mu_1 \mu_2 \mu_3}(p_1, p_2, p_3) = \left[-\frac{1}{24} C_A V^{\mu_1 \mu_2 \mu_3}(p_1, p_2, p_3) + \mathcal{O}(\varepsilon) \right] \frac{\alpha_s}{4 \pi \varepsilon}. \qquad (3.59)$$

The gluon-loop contribution (Fig. 3.22) has the colour factor $C_A/2$ (A.38). In the Feynman gauge,

$$\Lambda_g^{\mu_1 \mu_2 \mu_3}(p_1, p_2, p_3) = -i \frac{C_A}{2} g_0^2 \int \frac{d^d k}{(2\pi)^d} \frac{1}{(k^2)^3}$$
$$\times V^{\mu_1 \nu_1}{}_{\nu_3} \left(p_1, k + \frac{p_2 - p_1}{3}, -k - \frac{p_1 - p_3}{3} \right)$$
$$\times V^{\mu_3 \nu_3}{}_{\nu_2} \left(p_3, k + \frac{p_1 - p_3}{3}, -k - \frac{p_3 - p_2}{3} \right) \qquad (3.60)$$
$$\times V^{\mu_2 \nu_2}{}_{\nu_1} \left(p_2, k + \frac{p_3 - p_2}{3}, -k - \frac{p_2 - p_1}{3} \right).$$

Separating terms with two factors k in the numerator and averaging them

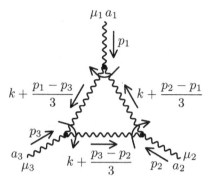

Fig. 3.22   The gluon-loop contribution to the three-gluon vertex

over k directions, we see that the numerator becomes

$$\frac{13}{4}k^2 V^{\mu_1\mu_2\mu_3}(p_1, p_2, p_3),$$

and the result is

$$\Lambda_g^{\mu_1\mu_2\mu_3}(p_1, p_2, p_3) = \left[\frac{13}{8}C_A V^{\mu_1\mu_2\mu_3}(p_1, p_2, p_3) + \mathcal{O}(\varepsilon)\right]\frac{\alpha_s}{4\pi\varepsilon}. \quad (3.61)$$

A much longer computation in an arbitrary covariant gauge produces

$$\Lambda_g^{\mu_1\mu_2\mu_3}(p_1, p_2, p_3) = \left[\frac{10 + 3a}{8}C_A V^{\mu_1\mu_2\mu_3}(p_1, p_2, p_3) + \mathcal{O}(\varepsilon)\right]\frac{\alpha_s}{4\pi\varepsilon}. \quad (3.62)$$

Finally, we consider the first diagram with a four-gluon vertex in Fig. 3.19 (the other two ones are obtained by simple substitutions). Decomposing the four-gluon vertex into our auxiliary-field notation (Fig. 3.4), we have two diagrams (Fig. 3.23). The first one has the symmetry factor $1/2$ and the colour factor C_A (A.30); the second one has the colour factor $C_A/2$ (A.38). In the Feynman gauge,

$$\Lambda_1^{\mu_1\mu_2\mu_3} = -i\frac{C_A}{2}g_0^2\int \frac{d^dk}{(2\pi)^d}\frac{V^{\mu_1}{}_{\nu_1\nu_2}\left(p_1, k - \frac{p_1}{2}, -k - \frac{p_1}{2}\right)}{\left(k - \frac{p_1}{2}\right)^2\left(k + \frac{p_1}{2}\right)^2} \quad (3.63)$$

$$\times (2g^{\mu_2\nu_2}g^{\mu_3\nu_1} - g^{\mu_2\nu_1}g^{\mu_3\nu_2} - g^{\mu_2\mu_3}g^{\nu_1\nu_2})$$

The three-gluon vertex $V^{\mu_1}{}_{\nu_1\nu_2}$ is linear in k; the vector integral with k in

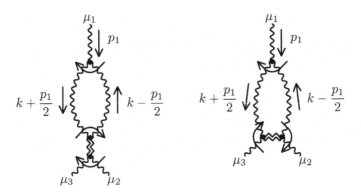

Fig. 3.23   The first diagram with a four-gluon vertex

the numerator vanishes, so, we can replace $k \to 0$ in it:

$$V^{\mu_1\nu_1\nu_2}\left(p_1, -\frac{p_1}{2}, -\frac{p_1}{2}\right) = \frac{3}{2}\left(p_1^{\nu_1}g^{\mu_1\nu_2} - p_1^{\nu_2}g^{\mu_1\nu_1}\right) .$$

Contracting this vertex with the last bracket, we obtain the numerator

$$\frac{9}{2}\left(p_1^{\mu_3}g^{\mu_1\mu_2} - p_1^{\mu_2}g^{\mu_1\mu_3}\right) .$$

Therefore, the ultraviolet divergence is

$$\Lambda_1^{\mu_1\mu_2\mu_3} = \frac{9}{4}C_A\left[1 + \mathcal{O}(\varepsilon)\right]\frac{\alpha_s}{4\pi\varepsilon}\left(p_1^{\mu_3}g^{\mu_1\mu_2} - p_1^{\mu_2}g^{\mu_1\mu_3}\right) . \qquad (3.64)$$

Adding the contributions of the other two diagrams (Fig. 3.19), which are obtained by cyclic permutations of 1, 2, 3, we arrive at

$$\Lambda_{\text{tot}}^{\mu_1\mu_2\mu_3} = \left[-\frac{9}{4}C_A V^{\mu_1\mu_2\mu_3}(p_1, p_2, p_3) + \mathcal{O}(\varepsilon)\right]\frac{\alpha_s}{4\pi\varepsilon} . \qquad (3.65)$$

It is not difficult to repeat this calculation in an arbitrary covariant gauge. All integrals in $\Lambda_1^{\mu_1\mu_2\mu_3}$ are one-loop propagator integrals (Sect. 1.5) depending on a single momentum p_1. Tensor integrals with up to 3 factors k in the numerator appear; they can be expanded in tensor structures constructed from p_1, their coefficients are obtained by solving linear systems from scalar integrals. The final result is

$$\Lambda_{\text{tot}}^{\mu_1\mu_2\mu_3} = \left[-\frac{3}{8}(7 - a)C_A V^{\mu_1\mu_2\mu_3}(p_1, p_2, p_3) + \mathcal{O}(\varepsilon)\right]\frac{\alpha_s}{4\pi\varepsilon} . \qquad (3.66)$$

When expressed via renormalized quantities, the three-gluon vertex (3.54) should be equal to

$$\Gamma^{\mu_1\mu_2\mu_3}(p_1,p_2,p_3) = Z_{3g}\Gamma_r^{\mu_1\mu_2\mu_3}(p_1,p_2,p_3)\,, \qquad (3.67)$$

where Z_{3g} is a minimal renormalization constant, and the renormalized vertex $\Gamma_r^{\mu_1\mu_2\mu_3}(p_1,p_2,p_3)$ is finite at $\varepsilon \to 0$. This means, in particular, that the ultraviolet-divergent part of $\Gamma^{\mu_1\mu_2\mu_3}(p_1,p_2,p_3)$ must have the tensor structure of the elementary vertex $V^{\mu_1\mu_2\mu_3}(p_1,p_2,p_3)$ (3.10). Collecting the results (3.57), (3.59), (3.62), and (3.66), we see that this requirement is fulfilled, and

$$Z_{3g} = 1 + \left(\frac{9a-17}{12}C_A + \frac{4}{3}T_F n_f\right)\frac{\alpha_s}{4\pi\varepsilon}\,. \qquad (3.68)$$

Physical matrix elements of three-gluon processes are obtained from the vertex $g_0\Gamma$ by multiplying by the external-line renormalization factor $Z_A^{1/2}$ for each external gluon leg (Sect. 2.7). They must be finite at $\varepsilon \to 0$:

$$g_0\Gamma Z_A^{3/2} = g\Gamma_r Z_\alpha^{1/2} Z_{3g} Z_A^{3/2} = \text{finite}\,.$$

Therefore,

$$Z_\alpha = Z_{3g}^{-2} Z_A^{-3}\,. \qquad (3.69)$$

Combining the results for Z_{3g} (3.68) and Z_A (3.28), we obtain the third derivation of Z_α (3.44).

3.9 Four-gluon vertex

In this Section, we shall calculate the ultraviolet-divergent part of the one-loop correction to the four-gluon vertex $-ig_0^2\Gamma_{\mu_1\mu_2\mu_3\mu_4}^{a_1a_2a_3a_4}$. This part has the same colour and tensor structure as the lowest-order vertex (Fig. 3.4). Therefore, we shall simplify our task by considering the contraction

$$\Gamma_{\mu_1\mu_2\mu_3\mu_4}^{a_1a_2a_3a_4}\,\delta^{a_1a_2}\delta^{a_3a_4}g^{\mu_1\mu_2}g^{\mu_3\mu_4} = 2d(d-1)C_0\,(1+\Lambda)\,,$$

$$C_0 = C_A N_g\,, \qquad N_g = \frac{C_F C_A}{2T_F^2}\,, \qquad (3.70)$$

where N_g is the number of gluon colours (A.24). In the one-loop approximation, Λ is given by the diagrams in Fig. 3.24. In some way, this problem is simpler than that for the three-gluon vertex, because here we can nullify

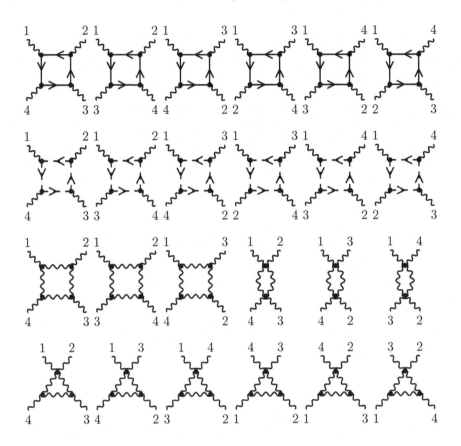

Fig. 3.24   One-loop diagrams for the four-gluon vertex

all external momenta (and introduce some infrared cutoff); all integrals are logarithmically divergent (1.29).

There are 6 quark-loop diagrams in Fig. 3.24. After taking the contraction (3.70), four of them are equal to the diagram in Fig. 3.25 contracted with $\delta_{12}\delta_{34} \equiv \delta^{a_1 a_2}\delta^{a_3 a_4} g^{\mu_1 \mu_2} g^{\mu_3 \mu_4}$, and two — with $\delta_{13}\delta_{24}$. The colour factor of the first contraction is

$$\frac{C_F C_A^2}{2T_F} = C_0 T_F \frac{C_F}{C_A} \,;$$

the "colourless" contribution is

$$-g_0^4 \int \frac{d^d k}{(2\pi)^d} \frac{\operatorname{Tr} \gamma^\mu \not{k} \gamma_\mu \not{k} \gamma^\nu \not{k} \gamma_\nu \not{k}}{(-k^2)^4} = -16 i g_0^2 \frac{\alpha_s}{4\pi\varepsilon}$$

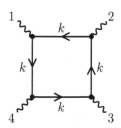

Fig. 3.25   Quark-loop diagram

(we retain only the ultraviolet $1/\varepsilon$ divergence, and hence may use 4-dimensional γ-matrix algebra). The colour factor of the second contraction is

$$-T_F C_F = C_0 T_F \left(\frac{C_F}{C_A} - \frac{1}{2} \right) ;$$

the "colourless" contribution is

$$-g_0^4 \int \frac{d^d k}{(2\pi)^d} \frac{\mathrm{Tr}\, \gamma^\mu \slashed{k} \gamma^\nu \slashed{k} \gamma_\mu \slashed{k} \gamma_\nu \slashed{k}}{(-k^2)^4} = 32 i g_0^2 \frac{\alpha_s}{4\pi\varepsilon} .$$

In the full contribution to Λ (3.70), C_F/C_A terms cancel:

$$\Lambda_q = \frac{4}{3} T_F n_f \frac{\alpha_s}{4\pi\varepsilon} . \tag{3.71}$$

There are 6 ghost-loop diagrams in Fig. 3.24. After taking the contraction (3.70), four of them are equal to the diagram in Fig. 3.26 contracted with $\delta_{12}\delta_{34}$, and two — with $\delta_{13}\delta_{24}$. The colour factor of the first contraction is $C_0 C_A$, and of the second one $C_0 C_A/2$. All "colourless" contributions

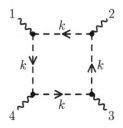

Fig. 3.26   Ghost-loop diagram

are the same:

$$-g_0^4 \int \frac{d^d k}{(2\pi)^d} \frac{1}{(-k^2)^2} = -ig_0^2 \frac{\alpha_s}{4\pi\varepsilon} .$$

The full contribution to Λ (3.70) is

$$\Lambda_c = \frac{5}{24} C_A \frac{\alpha_s}{4\pi\varepsilon} . \qquad (3.72)$$

There are three gluon-loop diagrams in Fig. 3.24. After taking the contraction (3.70), two of them are equal to the diagram in Fig. 3.27 contracted with $\delta_{12}\delta_{34}$, and one — with $\delta_{13}\delta_{24}$. The colour factor of the first contraction is $C_0 C_A$. Its "colourless" contribution in the Feynman gauge is

$$g_0^4 \int \frac{d^d k}{(2\pi)^d} \frac{1}{(-k^2)^4} V^{\mu_1\nu_1\nu_4}(0, k, -k) V_{\mu_1\nu_2\nu_1}(0, k, -k)$$
$$\times V^{\mu_2\nu_3\nu_2}(0, k, -k) V_{\mu_2\nu_4\nu_3}(0, k, -k) = 84 i g_0^2 \frac{\alpha_s}{4\pi\varepsilon} .$$

In an arbitrary covariant gauge, a slightly more lengthy calculation gives

$$12i(a^2 + 2a + 4) g_0^2 \frac{\alpha_s}{4\pi\varepsilon} .$$

The colour factor of the second contraction is $C_0 C_A / 2$. Its "colourless" contribution in the Feynman gauge is

$$g_0^4 \int \frac{d^d k}{(2\pi)^d} \frac{1}{(-k^2)^4} V^{\mu_1\nu_1\nu_4}(0, k, -k) V^{\mu_2\nu_2}{}_{\nu_1}(0, k, -k)$$
$$\times V_{\mu_1\nu_3\nu_2}(0, k, -k) V_{\mu_2\nu_4}{}^{\nu_3}(0, k, -k) = 54 i g_0^2 \frac{\alpha_s}{4\pi\varepsilon} ,$$

or in an arbitrary covariant gauge

$$6i(a^2 + 8) g_0^2 \frac{\alpha_s}{4\pi\varepsilon} .$$

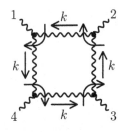

Fig. 3.27   Gluon-loop diagram

The full contribution to Λ (3.70) is

$$\Lambda_g = -\left(\frac{9}{8}a^2 + 2a + 5\right) C_A \frac{\alpha_s}{4\pi\varepsilon}. \tag{3.73}$$

There are 6 diagrams with a single four-gluon vertex in Fig. 3.24. After taking the contraction (3.70), four of them are equal. Each of these four diagrams becomes, in our auxiliary-field notation (Fig. 3.4), the sum of three diagrams: the second diagram in Fig. 3.28 contracted with $\delta_{14}\delta_{23}$; the same diagram contracted with $\delta_{13}\delta_{24}$; and the first diagram in Fig. 3.28 contracted with $\delta_{14}\delta_{23}$. The colour factors of these contractions are $C_0 C_A$, $C_0 C_A/2$, $C_0 C_A/2$. Their "colourless" contributions in the Feynman gauge are

$$T_2^{\mu_1\mu_2\mu_3\mu_4} g_{\mu_1\mu_4} g_{\mu_2\mu_3} = -27 i g_0^2 \frac{\alpha_s}{4\pi\varepsilon},$$

$$T_2^{\mu_1\mu_2\mu_3\mu_4} g_{\mu_1\mu_3} g_{\mu_2\mu_4} = -9 i g_0^2 \frac{\alpha_s}{4\pi\varepsilon},$$

$$T_1^{\mu_1\mu_2\mu_3\mu_4} g_{\mu_1\mu_4} g_{\mu_2\mu_3} = -18 i g_0^2 \frac{\alpha_s}{4\pi\varepsilon},$$

where the two tensor diagrams in Fig. 3.28 are

$$T_1^{\mu_1\mu_2\mu_3\mu_4} = g_0^4 \int \frac{d^d k}{(2\pi)^d} \frac{1}{(k^2)^3} \left(\delta_{\nu_2}^{\mu_1}\delta_{\nu_1}^{\mu_2} - \delta_{\nu_1}^{\mu_1}\delta_{\nu_2}^{\mu_2}\right)$$
$$\times V^{\mu_4\nu_3\nu_1}(0, -k, k) V^{\mu_3\nu_2}{}_{\nu_3}(0, -k, k),$$

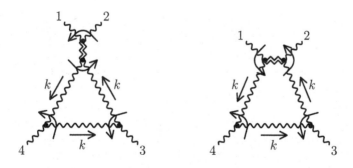

Fig. 3.28   Contributions with one four-gluon vertex

$$T_2^{\mu_1\mu_2\mu_3\mu_4} = g_0^4 \int \frac{d^dk}{(2\pi)^d} \frac{1}{(k^2)^3} \left(\delta_{\nu_2}^{\mu_1} \delta_{\nu_1}^{\mu_2} - g^{\mu_1\mu_2} g_{\nu_1\nu_2} \right)$$
$$\times V^{\mu_4\nu_3\nu_1}(0, -k, k) V^{\mu_3\mu_2}{}_{\nu_3}(0, -k, k) .$$

In an arbitrary covariant gauge, these three "colourless" contributions are

$$-3i(a+2)^2 g_0^2 \frac{\alpha_s}{4\pi\varepsilon}, \qquad -3i(a^2 - 2a + 4)g_0^2 \frac{\alpha_s}{4\pi\varepsilon}, \qquad -18iag_0^2 \frac{\alpha_s}{4\pi\varepsilon} .$$

The remaining two diagrams are equal. Each of them becomes, in our auxiliary-field notation (Fig. 3.4), the sum of three diagrams. One of them has zero colour factor; each of the other two is the contraction of the first tensor diagram in Fig. 3.28 with $\delta_{12}\delta_{34}$. Its colour factor is $C_0 C_A$, and the "colourless" contribution in the Feynman gauge is

$$T_1^{\mu_1\mu_2\mu_3\mu_4} g_{\mu_1\mu_2} g_{\mu_3\mu_4} = -54ig_0^2 \frac{\alpha_s}{4\pi\varepsilon} .$$

In an arbitrary covariant gauge, it is

$$-9i(a^2 + a + 4)g_0^2 \frac{\alpha_s}{4\pi\varepsilon} .$$

Collecting all these partial results together, we obtain the contribution of the diagrams with a single four-gluon vertex to Λ:

$$\Lambda_1 = \frac{9}{4}(a^2 + 2a + 4)C_A \frac{\alpha_s}{4\pi\varepsilon} . \qquad (3.74)$$

There are three diagrams with two four-gluon vertices in Fig. 3.24. After taking the contraction (3.70), using our auxiliary-field notation (Fig. 3.4), we obtain a number of contributions. Omitting those with zero colour factors, we have: the first diagram in Fig. 3.29 contracted with $\delta_{14}\delta_{23}$; the second diagram in Fig. 3.29 contracted with $\delta_{14}\delta_{23}$ multiplied by 4; the third diagram in Fig. 3.29 contracted with $\delta_{12}\delta_{34}$ multiplied by 2; the third diagram in Fig. 3.29 contracted with $\delta_{14}\delta_{23}$ multiplied by 2; the third diagram in Fig. 3.29 contracted with $\delta_{13}\delta_{24}$ multiplied by 2 (these factors count the numbers of contributions, with the account of their symmetry factors). The colour factors of these contributions are $C_0 C_A$, $C_0 C_A/2$, $C_0 C_A$, $C_0 C_A$, $C_0 C_A/2$. The "colourless" contributions in the Feynman

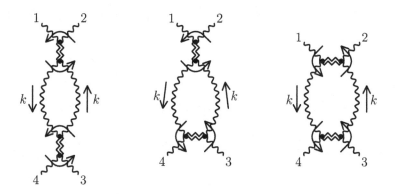

Fig. 3.29   Contributions with two four-gluon vertices

gauge are

$$T_1^{\mu_1\mu_2\mu_3\mu_4} g_{\mu_1\mu_4} g_{\mu_2\mu_3} = 24 i g_0^2 \frac{\alpha_s}{4\pi\varepsilon} ,$$

$$T_2^{\mu_1\mu_2\mu_3\mu_4} g_{\mu_1\mu_4} g_{\mu_2\mu_3} = 12 i g_0^2 \frac{\alpha_s}{4\pi\varepsilon} ,$$

$$T_3^{\mu_1\mu_2\mu_3\mu_4} g_{\mu_1\mu_2} g_{\mu_3\mu_4} = 36 i g_0^2 \frac{\alpha_s}{4\pi\varepsilon} ,$$

$$T_3^{\mu_1\mu_2\mu_3\mu_4} g_{\mu_1\mu_4} g_{\mu_2\mu_3} = 24 i g_0^2 \frac{\alpha_s}{4\pi\varepsilon} ,$$

$$T_3^{\mu_1\mu_2\mu_3\mu_4} g_{\mu_1\mu_3} g_{\mu_2\mu_4} = 12 i g_0^2 \frac{\alpha_s}{4\pi\varepsilon} ,$$

where the tensor diagrams of Fig. 3.29 are

$$T_1^{\mu_1\mu_2\mu_3\mu_4} = g_0^4 \int \frac{d^d k}{(2\pi)^d} \frac{1}{(k^2)^2} \left(g^{\mu_2\nu_1} g^{\mu_1\nu_2} - g^{\mu_2\nu_2} g^{\mu_1\nu_1} \right)$$
$$\times \left(\delta_{\nu_2}^{\mu_4} \delta_{\nu_1}^{\mu_3} - \delta_{\nu_1}^{\mu_4} \delta_{\nu_2}^{\mu_3} \right) ,$$

$$T_2^{\mu_1\mu_2\mu_3\mu_4} = g_0^4 \int \frac{d^d k}{(2\pi)^d} \frac{1}{(k^2)^2} \left(g^{\mu_2\nu_1} g^{\mu_1\nu_2} - g^{\mu_2\nu_2} g^{\mu_1\nu_1} \right)$$
$$\times \left(\delta_{\nu_2}^{\mu_4} \delta_{\nu_1}^{\mu_3} - g_{\nu_1\nu_2} g^{\mu_4\mu_3} \right) ,$$

$$T_3^{\mu_1\mu_2\mu_3\mu_4} = g_0^4 \int \frac{d^d k}{(2\pi)^d} \frac{1}{(k^2)^2} \left(g^{\mu_2\nu_1} g^{\mu_1\nu_2} - g^{\mu_1\mu_2} g^{\nu_1\nu_2} \right)$$
$$\times \left(\delta_{\nu_2}^{\mu_4} \delta_{\nu_1}^{\mu_3} - g_{\nu_1\nu_2} g^{\mu_4\mu_3} \right) .$$

In the arbitrary covariant gauge, the contributions are

$$12i(a+1)g_0^2 \frac{\alpha_s}{4\pi\varepsilon}, \quad 6i(a+1)g_0^2 \frac{\alpha_s}{4\pi\varepsilon}, \quad 9i(a^2+3)g_0^2 \frac{\alpha_s}{4\pi\varepsilon},$$

$$3i(a^2+2a+5)g_0^2 \frac{\alpha_s}{4\pi\varepsilon}, \quad 3i(a^2+3)g_0^2 \frac{\alpha_s}{4\pi\varepsilon}.$$

Combining these results, we obtain

$$\Lambda_2 = -\frac{3}{8}(3a^2 + 4a + 13)C_A \frac{\alpha_s}{4\pi\varepsilon}. \tag{3.75}$$

Collecting the results (3.71), (3.72), (3.73), (3.74), and (3.75), we obtain the four-gluon vertex renormalization constant

$$Z_{4g} = 1 + \left[\left(a - \frac{2}{3}\right)C_A + \frac{4}{3}T_F n_f\right]\frac{\alpha_s}{4\pi\varepsilon}. \tag{3.76}$$

The coupling constant renormalization can be obtained from this vertex, too:

$$Z_\alpha = Z_{4g}^{-1} Z_A^{-2}. \tag{3.77}$$

Substituting Z_A (3.28), we obtain the fourth derivation of Z_α (3.44).

Chapter 4

Two-loop corrections in QED and QCD

4.1 Massless propagator diagram

Let's consider the two-loop massless propagator diagram (Fig. 4.1),

$$
\int \frac{d^d k_1 \, d^d k_2}{D_1^{n_1} D_2^{n_2} D_3^{n_3} D_4^{n_4} D_5^{n_5}} = -\pi^d (-p^2)^{d - \sum n_i} G(n_1, n_2, n_3, n_4, n_5) \,,
$$
$$
D_1 = -(k_1 + p)^2 \,, \quad D_2 = -(k_2 + p)^2 \,, \quad D_3 = -k_1^2 \,, \quad D_4 = -k_2^2 \,, \tag{4.1}
$$
$$
D_5 = -(k_1 - k_2)^2 \,.
$$

The power of $-p^2$ is evident from dimensionality. Our aim is to calculate the dimensionless function $G(n_1, n_2, n_3, n_4, n_5)$. It is symmetric with respect to the interchanges $(1 \leftrightarrow 2, 3 \leftrightarrow 4)$ and $(1 \leftrightarrow 3, 2 \leftrightarrow 4)$. It vanishes when indices of two adjacent lines are non-positive integers, because then it contains a no-scale subdiagram.

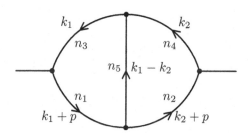

Fig. 4.1 Two-loop massless propagator diagram

When one of the indices is zero, the problem becomes trivial. If $n_5 = 0$,

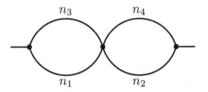

Fig. 4.2   Trivial case $n_5 = 0$

it is the product of two one-loop diagrams (Fig. 4.2):

$$G(n_1, n_2, n_3, n_4, 0) = G(n_1, n_3)G(n_2, n_4).$$
(4.2)

For integer n_i, the result is proportional to $G_1^2 = G(1, 1, 1, 1, 0)$, see (1.42).

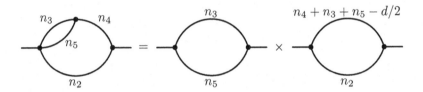

Fig. 4.3   Trivial case $n_1 = 0$

If $n_1 = 0$ (Fig. 4.3), then the inner loop gives

$$\frac{G(n_3, n_5)}{(-k_2^2)^{n_3 + n_5 - d/2}},$$

and hence

$$G(0, n_2, n_3, n_4, n_5) = G(n_3, n_5)G(n_2, n_4 + n_3 + n_5 - d/2).$$
(4.3)

The cases $n_2 = 0$, $n_3 = 0$, $n_4 = 0$ are given by the symmetric formulas. For integer n_i, the result is proportional to

$$G_2 = G(0, 1, 1, 0, 1) = G(1, 1)G(1, 2 - d/2)$$

$$= \frac{4g_2}{(d - 3)(d - 4)(3d - 8)(3d - 10)},$$
(4.4)

$$g_2 = \frac{\Gamma(1 + 2\varepsilon)\Gamma^3(1 - \varepsilon)}{\Gamma(1 - 3\varepsilon)}$$

(Fig. 4.4).

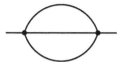

Fig. 4.4 Basis integral

But what can we do if all 5 indices are positive? We shall use integration by parts [Chetyrkin and Tkachov (1981)] — a powerful method based on the simple observation that integrals of full derivatives vanish. When applied to the integrand of (4.1), the derivative $\partial/\partial k_2$ acts as

$$\frac{\partial}{\partial k_2} \rightarrow \frac{n_2}{D_2} 2(k_2 + p) + \frac{n_4}{D_4} 2k_2 + \frac{n_5}{D_5} 2(k_2 - k_1).$$

Applying $(\partial/\partial k_2) \cdot (k_2 - k_1)$ to this integrand and using

$$(k_2 - k_1)^2 = -D_5, \quad 2k_2 \cdot (k_2 - k_1) = D_3 - D_4 - D_5,$$
$$2(k_2 + p) \cdot (k_2 - k_1) = D_1 - D_2 - D_5,$$

we see that this operation is equivalent to inserting

$$d - n_2 - n_4 - 2n_5 + \frac{n_2}{D_2}(D_1 - D_5) + \frac{n_4}{D_4}(D_3 - D_5)$$

under the integral sign in (4.1) (the term d comes from differentiating k_2). The resulting integral vanishes.

On the other hand, we can express it via G with shifted indices. Let's introduce the notation

$$\mathbf{1}^{\pm} G(n_1, n_2, n_3, n_4, n_5) = G(n_1 \pm 1, n_2, n_3, n_4, n_5), \qquad (4.5)$$

and similarly for $\mathbf{2}^{\pm}$, etc. Then the relation we have derived (it is called the triangle relation) takes the form

$$[d - n_2 - n_4 - 2n_5 + n_2 \mathbf{2}^{+}(\mathbf{1}^{-} - \mathbf{5}^{-}) + n_4 \mathbf{4}^{+}(\mathbf{3}^{-} - \mathbf{5}^{-})]G = 0. \qquad (4.6)$$

From (4.6) we obtain

$$G = \frac{n_2 \mathbf{2}^{+}(\mathbf{5}^{-} - \mathbf{1}^{-}) + n_4 \mathbf{4}^{+}(\mathbf{5}^{-} - \mathbf{3}^{-})}{d - n_2 - n_4 - 2n_5} G. \qquad (4.7)$$

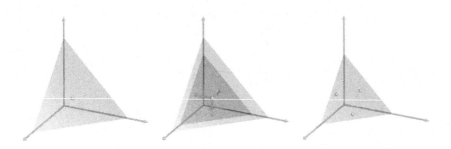

Fig. 4.5   Integration by parts

Each application of this relation reduces $n_1 + n_3 + n_5$ by 1 (Fig. 4.5). Therefore, sooner or later one of the indices n_1, n_3, n_5 will vanish, and we'll get a trivial case (4.2), (4.3), or symmetric to it.

Fig. 4.6   Two-loop propagator diagram

Let's summarize our achievements. There is one generic topology of two-loop propagator diagrams (Fig. 4.6). This means that all other topologies are reduced cases of this one, where some line (or lines) shrink(s) to a point, i.e., the corresponding index (-ices) vanish[1]. All Feynman integrals of this class, with any integer indices n_1, n_2, n_3, n_4, n_5, can be expressed, by a simple algorithm, as linear combinations of two basis integrals (Fig. 4.8), with coefficients being rational functions of d. The basis integrals are G_1^2

---

[1] At first sight it seems that the diagram with a self-energy insertion into a propagator (Fig. 4.7) is not a reduced case of Fig. 4.6. But this is not so: we can collect two identical denominators together, and then it obviously becomes Fig. 4.6 with one line shrunk.

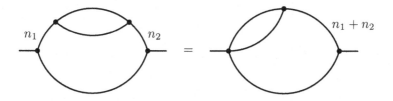

Fig. 4.7   Insertion into a propagator

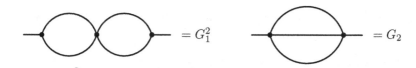

Fig. 4.8   Basis integrals

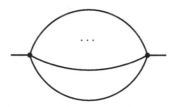

Fig. 4.9   Sunset diagrams

and G_2, where

$$
\begin{aligned}
G_n &= \frac{g_n}{\left(n+1-n\frac{d}{2}\right)_n \left((n+1)\frac{d}{2}-2n-1\right)_n}, \\
g_n &= \frac{\Gamma(1+n\varepsilon)\Gamma^{n+1}(1-\varepsilon)}{\Gamma(1-(n+1)\varepsilon)}
\end{aligned}
\tag{4.8}
$$

is the n-loop sunset diagram (Fig. 4.9).

Let's consider an example: Fig. 4.6 with all $n_i = 1$, i.e., $G(1,1,1,1,1)$.

Applying (4.7) to it, we obtain

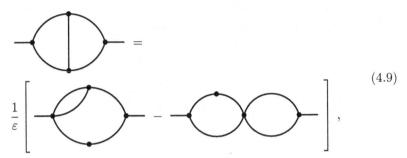

$$(4.9)$$

where the symmetry has been taken into account. Here a dot on a line means that this denominator is squared. We have

$$
\begin{aligned}
G(1,1,1,1,1) &= \frac{2}{d-4}\left[G(1,2,1,1,0) - G(0,2,1,1,1)\right] \\
&= \frac{2}{d-4}\left[G(1,1)G(2,1) - G(1,1)G(2,3-d/2)\right] \qquad (4.10) \\
&= \frac{2}{d-4}\left[\frac{G(2,1)}{G(1,1)}G_1^2 - \frac{G(2,-d/2+3)}{G(1,-d/2+2)}G_2\right],
\end{aligned}
$$

because

$$
G_1 = G(1,1), \quad G_2 = G(0,1,1,0,1) = G(1,1)G(1,-d/2+2).
$$

Using (1.41),

$$
\frac{G(2,3-d/2)}{G(1,2-d/2)} = -\frac{(3d-8)(3d-10)}{d-4},
$$

and we obtain

$$
\begin{aligned}
G(1,1,1,1,1) &= \frac{2}{d-4}\left[-(d-3)G_1^2 + \frac{(3d-8)(3d-10)}{d-4}G_2\right] \\
&= \frac{8(g_2 - g_1^2)}{(d-3)(d-4)^3}.
\end{aligned}
\qquad (4.11)
$$

It is easy to expand this result in ε using

$$
\frac{g_2}{g_1^2} = \frac{\Gamma(1+2\varepsilon)\Gamma^2(1-2\varepsilon)}{\Gamma^2(1+\varepsilon)\Gamma(1-\varepsilon)\Gamma(1-3\varepsilon)} = 1 - 6\zeta_3\varepsilon^3 + \cdots \qquad (4.12)
$$

We arrive at

$$
G(1,1,1,1,1) = 6\zeta_3 + \cdots \qquad (4.13)
$$

Three-loop massless propagator diagrams also can be calculated using integration by parts [Chetyrkin and Tkachov (1981)].

4.2 Photon self-energy

The photon self-energy at two loops is given by three diagrams (Fig. 4.10). It is gauge-invariant, because there are no off-shell charged external particles. Therefore, we may use any gauge; the calculation in the Feynman gauge $a_0 = 1$ is easiest.

Fig. 4.10   Two-loop photon self-energy

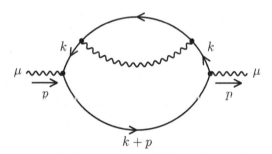

Fig. 4.11   The first contribution

The first two diagrams contribute equally; we only need to calculate one (Fig. 4.11) and double the result. This diagram contains one-loop electron self-energy subdiagram $-i\not{p}\Sigma_V$:

$$i\Pi_1{}_\mu^\mu = -\int \frac{d^d k}{(2\pi)^d} \operatorname{Tr} ie_0\gamma_\mu i\frac{\not{k}+\not{p}}{(k+p)^2} ie_0\gamma^\mu i\frac{\not{k}}{k^2}(-i)\not{k}\Sigma_V(k^2)i\frac{\not{k}}{k^2}.$$

Substituting the result (2.50) in the Feynman gauge $a_0 = 1$, we obtain

$$\Pi_{1\mu}^{\mu} = -i\frac{e_0^4}{(4\pi)^{d/2}}G_1\frac{d-2}{2}\int\frac{d^dk}{(2\pi)^d}\frac{\text{Tr}\,\gamma_\mu(\slashed{k}+\slashed{p})\gamma^\mu\slashed{k}}{[-(k+p)^2]\,(-k^2)^{1+\varepsilon}}$$

$$= i\frac{e_0^4}{(4\pi)^{d/2}}G_1(d-2)^2\int\frac{d^dk}{(2\pi)^d}\frac{2k\cdot(k+p)}{D_1 D_2^{1+\varepsilon}}\,.$$

Using the "multiplication table" (2.25) and omitting D_1 in the numerator, we have

$$\Pi_{1\mu}^{\mu} = -\frac{e_0^4(-p^2)^{1-2\varepsilon}}{(4\pi)^d}(d-2)^2 G(1,1)\left[G(1,1+\varepsilon) - G(1,\varepsilon)\right]$$

Using the property (1.41), we finally arrive at

$$\Pi_{1\mu}^{\mu} = -\frac{e_0^4(-p^2)^{1-2\varepsilon}}{(4\pi)^d}G_2\frac{2(d-2)^3}{d-4}\,. \tag{4.14}$$

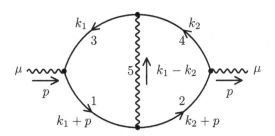

Fig. 4.12   The second contribution

The second contribution (Fig. 4.12) is more difficult. It is a truly two-loop diagram (Fig. 4.1):

$$i\Pi_{2\mu}^{\mu} = -\int\frac{d^dk_1}{(2\pi)^d}\frac{d^dk_2}{(2\pi)^d}\text{Tr}\,ie_0\gamma_\mu i\frac{\slashed{k}_2+\slashed{p}}{(k_2+p)^2}ie_0\gamma_\nu i\frac{\slashed{k}_1+\slashed{p}}{(k_1+p)^2}$$

$$\times\,ie_0\gamma^\mu i\frac{\slashed{k}_1}{k_1^2}ie_0\gamma^\nu i\frac{\slashed{k}_2}{k_2^2}\frac{-i}{(k_1-k_2)^2}\,,$$

or

$$\Pi_{2\mu}^{\mu} = -e_0^4\int\frac{d^dk_1}{(2\pi)^d}\frac{d^dk_2}{(2\pi)^d}\frac{N}{D_1 D_2 D_3 D_4 D_5}\,,$$

$$N = \text{Tr}\,\gamma_\mu(\slashed{k}_2+\slashed{p})\gamma_\nu(\slashed{k}_1+\slashed{p})\gamma^\mu\slashed{k}_1\gamma^\nu\slashed{k}_2\,.$$

According to (4.1), the "multiplication table" of the momenta is

$$p^2 = -1, \quad k_1^2 = -D_3, \quad k_2^2 = -D_4,$$

$$p \cdot k_1 = \frac{1}{2}(1 + D_3 - D_1), \quad p \cdot k_2 = \frac{1}{2}(1 + D_4 - D_2), \qquad (4.15)$$

$$k_1 \cdot k_2 = \frac{1}{2}(D_5 - D_3 - D_4).$$

All products $D_i D_j$ in the numerator can be omitted (produce vanishing integrals), except $D_1 D_4$ and $D_2 D_3$. Calculating the trace with this simplification, we obtain

$$N \Rightarrow 2(d-2)\Big[-(d-4)(D_1 D_4 + D_2 D_3)$$

$$+ 2(D_1 + D_2 + D_3 + D_4) - 2D_5^2 + (d-8)D_5 - 2\Big].$$

Therefore,

$$\Pi_{2\mu}^{\mu} = \frac{e_0^4(-p^2)^{1-2\varepsilon}}{(4\pi)^d}2(d-2)\Big[-2(d-4)G(0,1,1,0,1) + 8G(0,1,1,1,1)$$

$$- 2G(1,1,1,1,-1) + (d-8)G(1,1,1,1,0) - 2G(1,1,1,1,1)\Big].$$

Some integrals here are trivial:

$$G(1,1,1,1,0) = G_1^2, \quad G(0,1,1,0,1) = G_2,$$

or simple:

$$G(0,1,1,1,1) = G(1,1)G(1,3-d/2) = \frac{3d-8}{d-4}G_2$$

(here we used (1.41)). The only truly two-loop integral, $G(1,1,1,1,1)$, has been calculated in Sect. 4.1, and is given by (4.11). But what is $G(1,1,1,1,-1)$? This integral would factor into two one-loop ones (Fig. 4.2) if it had no numerator. In principle, it is possible to calculate such integrals using integration-by-parts recurrence relations. But we have not discussed the necessary methods. Therefore, we shall calculate this integral in a straightforward manner:

$$G(1,1,1,1,-1) = -\frac{1}{\pi^d}\int \frac{d^d k_1}{D_1 D_3}\frac{d^d k_2}{D_2 D_4}\big[-(k_1 - k_2)^2\big]$$

$$= -\frac{2}{\pi^d}\int \frac{k_1\,d^d k_1}{D_1 D_3} \cdot \int \frac{k_2\,d^d k_2}{D_2 D_4}.$$

Both vector integrals are directed along p; therefore, we may project them onto p:

$$G(1,1,1,1,-1) = -\frac{2}{\pi^d p^2}\left(\int \frac{k_1 \cdot p\, d^d k_1}{D_1 D_3}\right)^2 = -\frac{1}{2}G_1^2. \tag{4.16}$$

Collecting all this together, we get

$$\Pi_{2\,\mu}^{\mu} = \frac{e_0^4(-p^2)^{1-2\varepsilon}}{(4\pi)^d}2\frac{d-2}{d-4}\left[(d^2-7d+16)G_1^2 - 2\frac{d^3-6d^2+20d-32}{d-4}G_2\right]. \tag{4.17}$$

Separate contributions to $\Pi_{\mu\nu}(p)$ are not transverse, but their sum is, i.e., it has the structure (2.15). The full two-loop photon self-energy

$$\Pi_2(p^2) = -\frac{2\Pi_{1\,\mu}^{\mu} + \Pi_{2\,\mu}^{\mu}}{(d-1)(-p^2)}$$

is

$$\Pi_2(p^2) = \frac{e_0^4(-p^2)^{-2\varepsilon}}{(4\pi)^d}2\frac{d-2}{(d-1)(d-4)}$$

$$\times\left[-(d^2-7d+16)G_1^2 + 4\frac{(d-3)(d^2-4d+8)}{d-4}G_2\right]. \tag{4.18}$$

The one-loop result (Sect. 2.4) is given by (2.27).

4.3 Photon field renormalization

The transverse part of the photon propagator, up to e^4, has the form

$$p^2 D_\perp(p^2) = \frac{1}{1-\Pi(p^2)} = 1 + \frac{e_0^2(-p^2)^{-\varepsilon}}{(4\pi)^{d/2}}f_1(\varepsilon) + \frac{e_0^4(-p^2)^{-2\varepsilon}}{(4\pi)^d}f_2(\varepsilon) + \cdots \tag{4.19}$$

From (2.27) and (4.18), using (1.42) and (4.4), we have

$$\varepsilon f_1(\varepsilon) = -4\frac{1-\varepsilon}{(1-2\varepsilon)(3-2\varepsilon)}g_1,$$

$$\varepsilon^2 f_2(\varepsilon) = 4\frac{1-\varepsilon}{\varepsilon(3-2\varepsilon)}\left[\frac{6-3\varepsilon+4\varepsilon^2-4\varepsilon^3}{(1-2\varepsilon)^2(3-2\varepsilon)}g_1^2 - 2\frac{2-2\varepsilon+\varepsilon^2}{(1-3\varepsilon)(2-3\varepsilon)}g_2\right].$$

We can expand these functions in ε:

$$\varepsilon e^{\gamma\varepsilon}f_1(\varepsilon) = c_{10} + c_{11}\varepsilon + c_{12}\varepsilon^2 + \cdots$$

$$\varepsilon^2 e^{2\gamma\varepsilon}f_2(\varepsilon) = c_{20} + c_{21}\varepsilon + c_{22}\varepsilon^2 + \cdots$$

At one loop, using

$$g_1 = 1 - \frac{1}{2}\zeta_2\varepsilon^2 + \cdots$$

we obtain

$$\varepsilon e^{\gamma\varepsilon} f_1(\varepsilon) = -\frac{4}{3} - \frac{20}{9}\varepsilon + \left(\frac{2}{3}\zeta_2 - \frac{112}{27}\right)\varepsilon^2 + \cdots$$

At two loops, using (4.12), we see that $1/\varepsilon$ terms cancel, and

$$\varepsilon^2 e^{2\gamma\varepsilon} f_2(\varepsilon) = \frac{16}{9} + \frac{106}{27}\varepsilon + \left(16\zeta_3 - \frac{16}{9}\zeta_2 - \frac{7}{3}\right)\varepsilon^2 + \cdots$$

In order to re-express it via $\alpha(\mu)$, it is sufficient to use (2.10) with the one-loop renormalization constant (2.74), (2.75):

$$\begin{aligned}
p^2 D_\perp(p^2) = 1 &+ \frac{\alpha(\mu)}{4\pi\varepsilon} e^{-L\varepsilon}\left(1 - \beta_0\frac{\alpha}{4\pi\varepsilon} + \cdots\right)\varepsilon e^{\gamma\varepsilon} f_1(\varepsilon) \\
&+ \left(\frac{\alpha}{4\pi\varepsilon}\right)^2 e^{-2L\varepsilon}\varepsilon^2 e^{2\gamma\varepsilon} f_2(\varepsilon) + \cdots
\end{aligned}$$

(4.20)

where

$$L = \log\frac{-p^2}{\mu^2}$$

(indicating the argument of α in the α^2 term would be beyond our accuracy). Substituting the ε expansions, we obtain at $L = 0$ (i.e. $\mu^2 = -p^2$)

$$\begin{aligned}
p^2 D_\perp(p^2) = 1 &+ \frac{\alpha(\mu)}{4\pi\varepsilon}(c_{10} + c_{11}\varepsilon + c_{12}\varepsilon^2 + \cdots) \\
&+ \left(\frac{\alpha}{4\pi\varepsilon}\right)^2 \left[c_{20} + c_{21}\varepsilon + c_{22}\varepsilon^2 + \cdots - \beta_0(c_{10} + c_{11}\varepsilon + c_{12}\varepsilon^2 + \cdots)\right] + \cdots
\end{aligned}$$

This should be equal to $Z_A(\alpha(\mu))p^2 D_\perp^r(p^2; \mu)$, where

$$Z_A(\alpha) = 1 + \frac{\alpha}{4\pi\varepsilon}z_1 + \left(\frac{\alpha}{4\pi\varepsilon}\right)^2 (z_{20} + z_{21}\varepsilon) + \cdots \qquad (4.21)$$

and

$$p^2 D_\perp^r(p^2; \mu) = 1 + \frac{\alpha(\mu)}{4\pi}(r_1 + r_{11}\varepsilon + \cdots) + \left(\frac{\alpha}{4\pi}\right)^2 (r_2 + \cdots) + \cdots \qquad (4.22)$$

Equating α terms, we obtain

$$z_1 = c_{10}, \quad r_1 = c_{11}, \quad r_{11} = c_{12}. \qquad (4.23)$$

Equating α^2 terms, we obtain

$$z_{20} = c_{20} - \beta_0 c_{10} , \quad z_{21} = c_{21} - (c_{10} + \beta_0)c_{11} , \quad r_2 = c_{22} - (c_{10} + \beta_0)c_{12} . \tag{4.24}$$

Therefore, we arrive at the photon field renormalization constant

$$Z_A(\alpha) = 1 - \frac{4}{3}\frac{\alpha}{4\pi\varepsilon} - 2\varepsilon \left(\frac{\alpha}{4\pi\varepsilon}\right)^2 + \cdots \tag{4.25}$$

and the renormalized photon propagator at $\mu^2 = -p^2$ (and $\varepsilon = 0$)

$$p^2 D_\perp^r(p^2; \mu^2 = -p^2) = 1 - \frac{20}{9}\frac{\alpha(\mu)}{4\pi} + \left(16\zeta_3 - \frac{55}{3}\right)\left(\frac{\alpha}{4\pi}\right)^2 + \cdots \tag{4.26}$$

The anomalous dimension (2.35) of the photon field is

$$
\begin{aligned}
\gamma_A &= \frac{d\log Z_A}{d\log\mu} = \frac{d}{d\log\mu}\left[z_1\frac{\alpha}{4\pi\varepsilon} + \left(z_{20} - \frac{1}{2}z_1^2 + z_{21}\varepsilon\right)\left(\frac{\alpha}{4\pi\varepsilon}\right)^2\right] \\
&= z_1\frac{\alpha}{4\pi\varepsilon}\left(-2\varepsilon - 2\beta_0\frac{\alpha}{4\pi}\right) + 2\left(z_{20} - \frac{1}{2}z_1^2 + z_{21}\varepsilon\right)\left(\frac{\alpha}{4\pi\varepsilon}\right)^2(-2\varepsilon) \\
&= -2z_1\frac{\alpha}{4\pi} - 4\left[\left(z_{20} - \frac{1}{2}z_1^2 + \frac{1}{2}\beta_0 z_1\right)\frac{1}{\varepsilon} + z_{21}\right]\left(\frac{\alpha}{4\pi}\right)^2 .
\end{aligned}
\tag{4.27}
$$

It must be finite at $\varepsilon \to 0$. Therefore, z_{20}, the coefficient of $1/\varepsilon^2$ in the two-loop term in Z_A, cannot be arbitrary. It must satisfy

$$z_{20} = \frac{1}{2}z_1(z_1 - \beta_0) . \tag{4.28}$$

In other words, c_{20}, the coefficient of $1/\varepsilon^2$ in the two-loop term in $D_\perp(p^2)$ (4.19), must satisfy

$$c_{20} = \frac{1}{2}c_{10}(c_{10} + \beta_0) . \tag{4.29}$$

Then

$$\gamma_{A0} = -2z_1 , \quad \gamma_{A1} = -4z_{21} ,$$

i.e., the coefficients in the anomalous dimension are determined by the coefficients of $1/\varepsilon$ in Z_A. Therefore, Z_A must have the form

$$Z_A = 1 - \frac{1}{2}\gamma_{A0}\frac{\alpha}{4\pi\varepsilon} + \frac{1}{8}\left[\gamma_{A0}(\gamma_{A0} + 2\beta_0) - 2\gamma_{A1}\varepsilon\right]\left(\frac{\alpha}{4\pi\varepsilon}\right)^2 + \cdots \tag{4.30}$$

This is indeed so (see (2.76) for β_0), and we obtain

$$\gamma_A(\alpha) = \frac{8}{3}\frac{\alpha}{4\pi} + 8\left(\frac{\alpha}{4\pi}\right)^2 + \cdots \tag{4.31}$$

This can also be understood in a slightly different way. The information contained in Z_A is equivalent to that in $\gamma_A(\alpha)$. This renormalization constant is gauge-invariant, because $\Pi(p^2)$ is gauge-invariant in QED (there are no off-shell external charged particles in it; this is not so in QCD, where gluons are "charged"). Therefore, Z_A depends on μ only via $\alpha(\mu)$ (there is no $a(\mu)$ in it). Dividing (2.35) by (2.72), we obtain

$$\frac{d\log Z_A}{d\log\alpha} = -\frac{1}{2}\frac{\gamma_A(\alpha)}{\varepsilon + \beta(\alpha)} = -\frac{\gamma_A(\alpha)}{2\varepsilon} + \frac{\beta(\alpha)\gamma_A(\alpha)}{2\varepsilon^2} + \cdots \tag{4.32}$$

Any minimal (2.8) renormalization constant can be represented as

$$Z_A = \exp\left(\frac{Z_1}{\varepsilon} + \frac{Z_2}{\varepsilon^2} + \cdots\right), \tag{4.33}$$

where Z_1 starts from the order α, Z_2 — from α^2, and so on. Then

$$\frac{dZ_1}{d\log\alpha} = -\frac{1}{2}\gamma_A(\alpha), \qquad \frac{dZ_2}{d\log\alpha} = \frac{1}{2}\beta(\alpha)\gamma_A(\alpha), \ldots$$

and

$$\begin{aligned}
Z_1 &= -\frac{1}{2}\int_0^\alpha \gamma_A(\alpha)\frac{d\alpha}{\alpha} = -\frac{1}{2}\gamma_{A0}\frac{\alpha}{4\pi} - \frac{1}{4}\gamma_{A1}\left(\frac{\alpha}{4\pi}\right)^2 - \cdots \\
Z_2 &= \frac{1}{2}\int_0^\alpha \beta(\alpha)\gamma_A(\alpha)\frac{d\alpha}{\alpha} = \frac{1}{4}\beta_0\gamma_{A0}\left(\frac{\alpha}{4\pi}\right)^2 + \cdots
\end{aligned} \tag{4.34}$$

$\cdots$

One can obtain $\gamma_A(\alpha)$ from Z_1, the coefficient of $1/\varepsilon$ in Z_A, and vice versa. Higher poles ($1/\varepsilon^2$, $1/\varepsilon^3$, ...) contain no new information: at each order in α, their coefficients (Z_2, Z_3, ...) can be reconstructed from lower-loop results. Up to two loops, this gives us (4.30).

Now let's return to (4.20) with arbitrary L. It should be equal to the product of (4.21) and (4.22). Equating α terms, we obtain

$$z_1 = c_{10}, \quad r_1(L) = c_{11} - c_{10}L, \quad r_{11}(L) = c_{12} - c_{11}L + c_{10}\frac{L^2}{2}.$$

Equating α^2 terms, we obtain

$$z_{20} = c_{20} - \beta_0 c_{10}\,, \quad z_{21} = c_{21} - 2c_{20}L - (c_{10} + \beta_0)(c_{11} - c_{10}L)\,,$$

$$r_2(L) = c_{22} - 2c_{21}L + 2c_{20}L^2 - (c_{10} + \beta_0)\left(c_{12} - c_{11}L + c_{10}\frac{L^2}{2}\right)\,.$$

But the renormalization constant Z_A cannot depend on kinematics of a specific process, i.e., on L. Therefore, terms with L in z_{21} must cancel. This is ensured by the consistency condition (4.29). The renormalization constant is given by (4.30), and the renormalized propagator at $\varepsilon = 0$ — by

$$p^2 D_\perp^r(p^2; \mu) = 1 + \frac{\alpha(\mu)}{4\pi}\left(r_1(0) + \frac{1}{2}\gamma_{A0}L\right)$$
$$+ \left(\frac{\alpha}{4\pi}\right)^2\left[r_2(0) + \frac{1}{2}\left(\gamma_{A1} + r_1(0)\left(\gamma_{A0} - 2\beta_0\right)\right)L + \frac{1}{8}\gamma_{A0}(\gamma_{A0} - 2\beta_0)L^2\right]$$
$$+ \cdots$$

$$(4.35)$$

This result can be also obtained by solving the RG equation (2.37). Taking into account

$$\frac{d\alpha}{dL} = \beta(\alpha)\alpha\,,$$

we find that the coefficients obey the following equations

$$\frac{dr_1(L)}{dL} = \frac{1}{2}\gamma_0\,, \quad \frac{dr_2(L)}{dL} = \frac{1}{2}\gamma_1 + \left(\frac{1}{2}\gamma_0 - \beta_0\right)r_1(L)\,.$$

Solving them, we reproduce (4.35).

Substituting the coefficients, we obtain

$$p^2 D_\perp^r(p^2; \mu) = 1 + \frac{\alpha(\mu)}{4\pi}\left(\frac{4}{3}L - \frac{20}{9}\right)$$
$$+ \left(\frac{\alpha(\mu)}{4\pi}\right)^2\left(\frac{16}{9}L^2 - \frac{52}{27}L + 10\zeta_3 - \frac{55}{3}\right) + \cdots$$

$$(4.36)$$

This is a typical example of perturbative series for a quantity with a single energy scale $(-p^2)$. In principle, we can choose the renormalization scale μ arbitrarily: physical results don't depend on it. However, if we choose it far away from the energy scale of the process, the coefficients in the series contain powers of the large logarithm L, and hence are large. Therefore, truncating the series after some term produces large errors. It is better to choose the renormalization scale of order of the characteristic energy scale, then $|L| \lesssim 1$. The coefficients contain no large logarithm and are just numbers (one hopes, of order 1). The convergence is better.

For example, when describing QED processes at LEP, at energies $\sim m_W$, it would be a very poor idea to use the low-energy α with $\mu \sim m_e$: coefficients of perturbative series would contain powers of a huge logarithm $\log(m_W/m_e)$. Using $\alpha(m_W)$, which is about 7% larger, makes the behaviour of perturbative series much better.

This renormalization-group improvement works for all processes with a single characteristic energy scale. If there are several widely separated scales, no universal method exists. For some classes of processes, there are some specific methods, such as, e.g., factorization. But this depends on the process under consideration very much.

4.4 Charge renormalization

In QED, it is enough to know Z_A to obtain charge renormalization (2.66). With two-loop accuracy, we have from (4.25)

$$Z_\alpha = Z_A^{-1} = 1 + \frac{4}{3}\frac{\alpha}{4\pi\varepsilon} + \left(\frac{16}{9} + 2\varepsilon\right)\left(\frac{\alpha}{4\pi\varepsilon}\right)^2. \tag{4.37}$$

The β-function (2.73) is simply

$$\beta(\alpha) = -\frac{1}{2}\gamma_A(\alpha), \tag{4.38}$$

or (see (4.31))

$$\beta(\alpha) = -\frac{4}{3}\frac{\alpha}{4\pi} - 4\left(\frac{\alpha}{4\pi}\right)^2 + \cdots \tag{4.39}$$

The information contained in Z_α is equivalent to that in $\beta(\alpha)$. Dividing (2.73) by (2.72), we have

$$\frac{d\log Z_\alpha}{d\log\alpha} = -\frac{\beta(\alpha)}{\varepsilon + \beta(\alpha)} = -\frac{\beta(\alpha)}{\varepsilon} + \frac{\beta^2(\alpha)}{\varepsilon^2} + \cdots \tag{4.40}$$

Writing Z_α in the form (4.33), we have

$$\frac{dZ_1}{d\log\alpha} = -\beta(\alpha), \quad \frac{dZ_2}{d\log\alpha} = \beta^2(\alpha), \ldots$$

and

$$Z_1 = -\int_0^\alpha \beta(\alpha)\frac{d\alpha}{\alpha} = -\beta_0\frac{\alpha}{4\pi} - \frac{1}{2}\beta_1\left(\frac{\alpha}{4\pi}\right)^2 - \cdots$$

$$Z_2 = \int_0^\alpha \beta^2(\alpha)\frac{d\alpha}{\alpha} = \frac{1}{2}\beta_0^2\left(\frac{\alpha}{4\pi}\right)^2 + \cdots \tag{4.41}$$

$$\cdots$$

One can obtain $\beta(\alpha)$ from Z_1, the coefficient of $1/\varepsilon$ in Z_α, and vice versa. Higher poles ($1/\varepsilon^2$, $1/\varepsilon^3$, ...) contain no new information: at each order in α, their coefficients (Z_2, Z_3, ...) can be reconstructed from lower-loop results. Up to two loops,

$$Z_\alpha = 1 - \beta_0\frac{\alpha}{4\pi\varepsilon} + \left(\beta_0^2 - \frac{1}{2}\beta_1\varepsilon\right)\left(\frac{\alpha}{4\pi\varepsilon}\right)^2 + \cdots \tag{4.42}$$

Comparing this with (4.37), we again obtain (4.39).

We shall later need $\beta(\alpha)$ for QED with n_f massless lepton fields (e.g., electrons and muons at energies $\gg m_\mu$). The photon self-energy at one loop (2.27) (Fig. 2.4) and at two loops (4.18) (Fig. 4.10) gets the factor n_f (at three loops, there are both n_f and n_f^2 terms). If $1 - \Pi(p^2)$ is expressed via the renormalized $\alpha(\mu)$, it should be equal to

$$1 - \Pi(p^2) = Z_\alpha(\alpha(\mu))\left[p^2 D_\perp^r(p^2;\mu)\right]^{-1},$$

where $Z_\alpha(\alpha)$ is minimal (2.8), and the second factor is finite at $\varepsilon \to 0$. Now we shall follow the derivation in Sect. 4.3. At one loop (4.23), z_1 gets the factor n_f, because c_{10} gets it (from $\Pi_1(p^2)$). Therefore, β_0 becomes n_f times larger. At two loops (4.24), z_{21} also simply gets the factor n_f from c_{21}, because $c_{10} + \beta_0 = z_1 + \beta_0 = 0$ for Z_α (4.42). Therefore,

$$\beta(\alpha) = -\frac{4}{3}n_f\frac{\alpha}{4\pi} - 4n_f\left(\frac{\alpha}{4\pi}\right)^2 + \cdots \tag{4.43}$$

At three loops, both n_f and n_f^2 terms appear. The photon field anomalous dimension is, from (4.38),

$$\gamma_A(\alpha) = \frac{8}{3}n_f\frac{\alpha}{4\pi} + 8n_f\left(\frac{\alpha}{4\pi}\right)^2 + \cdots \tag{4.44}$$

It is easy to write down the renormalization constants Z_α (4.42) and Z_A (4.30).

4.5 Electron self-energy

The electron self-energy at two loops is given by three diagrams (Fig. 4.13).

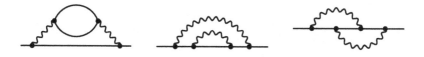

Fig. 4.13   Two-loop electron self-energy

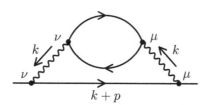

Fig. 4.14   Photon self-energy insertion

The first diagram (Fig. 4.14) contains one-loop photon self-energy sub-diagram $i \left(k^2 g_{\mu\nu} - k_\mu k_\nu \right) \Pi(k^2)$. This tensor is transverse, and hence longitudinal parts of the photon propagators drop out. Therefore, this diagram is gauge-invariant:

$$-i\Sigma_{V1}\slashed{p} = \int \frac{d^d k}{(2\pi)^d} ie_0\gamma^\mu i\frac{\slashed{k}+\slashed{p}}{(k+p)^2} ie_0\gamma^\nu \left(\frac{-i}{k^2} \right)^2 i \left(k^2 g_{\mu\nu} - k_\mu k_\nu \right) \Pi(k^2) .$$

We shall set $p^2 = -1$; the power of $-p^2$ will be restored in the result by dimensionality. Taking $\frac{1}{4} \operatorname{Tr} \slashed{p}$ of both sides and substituting the result (2.27), we obtain

$$\Sigma_{V1} = i\frac{e_0^4}{(4\pi)^{d/2}} \frac{d-2}{d-1} G_1 \int \frac{d^d k}{(2\pi)^d} \frac{N}{D_1 D_2^{2+\varepsilon}} ,$$

$$N = \frac{1}{2} \operatorname{Tr} \slashed{p}\gamma^\mu(\slashed{k}+\slashed{p})\gamma^\nu \cdot \left(k^2 g_{\mu\nu} - k_\mu k_\nu \right) .$$

Using the "multiplication table" (2.25) and omitting D_1 in the numerator, we have

$$N \Rightarrow (d-2)D_2^2 - (d-3)D_2 - 1 ,$$

and

$$\Sigma_{V1} = \frac{e_0^4}{(4\pi)^d} \frac{d-2}{d-1} G_1 \left[G(1, 2+\varepsilon) + (d-3)G(1, 1+\varepsilon) - (d-2)G(1, \varepsilon) \right].$$

Finally, using (1.41) and restoring the power of $-p^2$, we arrive at

$$\Sigma_{V1} = \frac{e_0^4(-p^2)^{-2\varepsilon}}{(4\pi)^d} G_2 \frac{2(d-2)^2}{d-6}. \qquad (4.45)$$

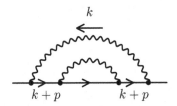

Fig. 4.15 Electron self-energy insertion

The second diagram (Fig. 4.15) contains one-loop electron self-energy subdiagram $-i\Sigma_V((k+p)^2)(\slashed{k}+\slashed{p})$. For simplicity, we shall calculate it in the Feynman gauge $a_0 = 1$:

$$-i\Sigma_{V2}\slashed{p} =$$

$$\int \frac{d^d k}{(2\pi)^d} ie_0\gamma^\mu i\frac{\slashed{k}+\slashed{p}}{(k+p)^2}(-i)\Sigma_V((k+p)^2)(\slashed{k}+\slashed{p})i\frac{\slashed{k}+\slashed{p}}{(k+p)^2}ie_0\gamma_\mu\frac{-i}{k^2}.$$

Taking $\frac{1}{4}\operatorname{Tr}\slashed{p}$ (with $p^2 = -1$) and substituting the result (2.50), we obtain

$$\Sigma_{V2} = -i\frac{e_0^4}{(4\pi)^{d/2}} \frac{d-2}{4} G_1 \int \frac{d^d k}{(2\pi)^d} \frac{N}{D_1^{1+\varepsilon} D_2}, \quad N = \frac{1}{2}\operatorname{Tr}\slashed{p}\gamma^\mu(\slashed{k}+\slashed{p})\gamma_\mu.$$

Using (2.25) and omitting D_2 in the numerator, we have $N \Rightarrow (d-2)(D_1 + 1)$, or

$$\Sigma_{V2} = \frac{e_0^4}{(4\pi)^d} \frac{(d-2)^2}{4} G_1 \left[G(1, 1+\varepsilon) + G(1, \varepsilon) \right].$$

Finally, we arrive at

$$\Sigma_{V2} = \frac{e_0^4(-p^2)^{-2\varepsilon}}{(4\pi)^d} G_2 \frac{(d-2)^2(d-3)}{d-4}. \qquad (4.46)$$

The result in the arbitrary covariant gauge is very simple:

$$\Sigma_{V2} = \frac{e_0^4(-p^2)^{-2\varepsilon}}{(4\pi)^d} G_2 \frac{(d-2)^2(d-3)}{d-4} a_0^2. \tag{4.47}$$

The reason is following. It is not difficult to show (using (1.41)) that the diagram of Fig. 2.6 (see (2.46)) with the denominator D_1 raised to an arbitrary power n instead of 1 is proportional to a_0 (prove this!). This means that in order to calculate this diagram with an arbitrary insertion(s) into the electron line, we may take the upper photon propagator in the Feynman gauge, and then multiply the result by a_0.

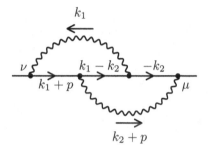

Fig. 4.16   Truly two-loop diagram

The third diagram (Fig. 4.16) is truly two-loop; it has the topology of Fig. 4.1. In the Feynman gauge,

$$-i\Sigma_{V3}\slashed{p} = \int \frac{d^d k_1}{(2\pi)^d} \frac{d^d k_2}{(2\pi)^d} ie_0\gamma^\mu i\frac{-\slashed{k}_2}{k_2^2} ie_0\gamma^\nu i\frac{\slashed{k}_1 - \slashed{k}_2}{(k_1 - k_2)^2} ie_0\gamma_\mu i\frac{\slashed{k}_1 + \slashed{p}}{(k_1 + p)^2} ie_0\gamma_\nu$$
$$\times \frac{-i}{k_1^2} \frac{-i}{(k_2 + p)^2},$$

or

$$\Sigma_{V3} = e_0^4 \int \frac{d^d k_1}{(2\pi)^d} \frac{d^d k_2}{(2\pi)^d} \frac{N}{D_1 D_2 D_3 D_4 D_5},$$
$$N = \frac{1}{4} \mathrm{Tr}\, \slashed{p}\gamma^\mu \slashed{k}_2 \gamma^\nu (\slashed{k}_1 - \slashed{k}_2)\gamma_\mu(\slashed{k}_1 + \slashed{p})\gamma_\nu.$$

Using the "multiplication table" (4.15) and omitting all products $D_i D_j$

except $D_1 D_4$ and $D_2 D_3$, we have

$$N \Rightarrow \frac{d-2}{2} \Big[(d-4) D_1 D_4 - (d-8) D_2 D_3$$
$$+ 2(D_1 - D_2 - D_3 + D_4) + 2(D_2^2 + D_3^2) + (d-4) D_5 \Big].$$

Therefore,

$$\Sigma_{V3} = -\frac{e_0^4 (-p^2)^{-2\varepsilon}}{(4\pi)^d} \frac{d-2}{2}$$
$$\times \Big[4G(0,1,1,0,1) + 4G(1,-1,1,1,1) + (d-4) G(1,1,1,1,0) \Big].$$

What is $G(1,-1,1,1,1)$? It would be trivial to calculate first the inner loop integral $(d^d k_2)$ and then the outer one $(d^d k_1)$, if there were no D_2 in the numerator. In principle, it is possible to calculate such integrals using integration-by-parts recurrence relations. But we have not discussed the necessary methods. Therefore, we shall calculate this integral in a straightforward manner:

$$G(1,-1,1,1,1) = -\frac{1}{\pi^d} \int \frac{d^d k_1}{D_1 D_3} \int \frac{d^d k_2}{D_4 D_5} \left[-(k_2 + p)^2 \right]$$
$$= -\frac{1}{\pi^d} \int \frac{d^d k_1}{D_1 D_3} \int \frac{d^d k_2}{D_4 D_5} (1 - 2p \cdot k_2) .$$

In the second term, the inner vector integral is directed along k_1, and we may substitute

$$k_2 \to \frac{k_2 \cdot k_1}{k_1^2} k_1 \to \frac{1}{2} k_1$$

(the terms with D_4 or D_5 in the numerator give vanishing integrals). Therefore,

$$G(1,-1,1,1,1) = -\frac{1}{\pi^d} \int \frac{d^d k_1}{D_1 D_3} (1 - p \cdot k_1) \int \frac{d^d k_2}{D_4 D_5}$$
$$= -\frac{1}{\pi^d} \int \frac{d^d k_1}{D_1 D_3} \frac{1}{2} (1 - D_3) \int \frac{d^d k_2}{D_4 D_5}$$
$$= \frac{1}{2} \left[G(1,0,1,1,1) - G(1,0,0,1,1) \right] .$$

Using the trivial integrals

$$G(1,0,0,1,1) = G_2 , \quad G(1,0,1,1,1) = \frac{3d-8}{d-4} G_2 ,$$

we obtain

$$G(1, -1, 1, 1, 1) = \frac{d-2}{d-4} G_2 .$$
(4.48)

Substituting the integrals, we obtain the Feynman-gauge result

$$\Sigma_{V3} = -\frac{e_0^4 (-p^2)^{-2\varepsilon}}{(4\pi)^d} \frac{d-2}{2} \left[(d-4)G_1^2 + 8 \frac{d-3}{d-4} G_2 \right] .$$
(4.49)

A similar, but more lengthy calculation in the arbitrary covariant gauge yields

$$\Sigma_{V3} = -\frac{e_0^4 (-p^2)^{-2\varepsilon}}{(4\pi)^d} \frac{d-2}{4}$$
$$\times \left[((d-2)a_0^2 + d - 6)G_1^2 - 2 \frac{d-3}{d-4} ((d-4)a_0^2 - d - 4) \right] .$$
(4.50)

We shall consider a more general case of QED with n_f massless lepton fields; the results for the usual QED can be easily obtained by setting $n_f = 1$. Then the diagram of Fig. 4.14 gives (4.45) with the extra factor n_f; the diagram of Fig. 4.15 gives (4.47), and that of Fig. 4.16 gives (4.50). The complete result for the two-loop term in $\Sigma_V(p^2)$ is

$$\Sigma_{2V}(p^2) = \frac{e_0^4 (-p^2)^{-2\varepsilon}}{(4\pi)^d} (d-2) \left[2 \frac{d-2}{d-6} G_2 n_f - \frac{1}{4} ((d-2)a_0^2 + d - 6)G_1^2 \right.$$
$$\left. + \frac{1}{2} \frac{d-3}{d-4} ((3d-8)a_0^2 - d - 4)G_2 \right] .$$
(4.51)

4.6 Electron field renormalization

Now we proceed as in Sect. 4.3. The electron propagator $\not{p}S(p)$ has the form similar to (4.19), but now the coefficients are gauge-dependent:

$$f_1(\varepsilon) = f_1'(\varepsilon) + f_1''(\varepsilon)a_0 .$$

From (2.50) and (4.51), using (1.42) and (4.4), we have

$$
\varepsilon f_1(\varepsilon) = -\frac{1-\varepsilon}{1-2\varepsilon} g_1 a_0 \,,
$$

$$
\varepsilon^2 f_2(\varepsilon) = (1-\varepsilon) \left[\frac{2\varepsilon(1-\varepsilon)}{(1+\varepsilon)(1-2\varepsilon)(1-3\varepsilon)(2-3\varepsilon)} g_2 n_f \right.
$$

$$
\left. + \frac{1+\varepsilon}{(1-2\varepsilon)^2} g_1^2 - \frac{4-\varepsilon-(2-3\varepsilon)a_0^2}{2(1-3\varepsilon)(2-3\varepsilon)} g_2 \right] .
$$

We can expand these functions in ε:

$$
\varepsilon e^{\gamma\varepsilon} f_1'(\varepsilon) = c_{10}' + c_{11}'\varepsilon + c_{12}'\varepsilon^2 + \cdots ,
$$

$$
\varepsilon e^{\gamma\varepsilon} f_1''(\varepsilon) = c_{10}'' + c_{11}''\varepsilon + c_{12}''\varepsilon^2 + \cdots
$$

$$
\varepsilon^2 e^{2\gamma\varepsilon} f_2(\varepsilon) = c_{20} + c_{21}\varepsilon + c_{22}\varepsilon^2 + \cdots
$$

In our particular case, $f_1'(\varepsilon) = 0$,

$$
\varepsilon e^{\gamma\varepsilon} f_1''(\varepsilon) = -1 - \varepsilon - \left(2 - \frac{1}{2}\zeta_2\right)\varepsilon^2 + \cdots
$$

$$
\varepsilon^2 e^{2\gamma\varepsilon} f_2(\varepsilon) = \frac{a_0^2}{2} + \left(n_f + a^2 + \frac{3}{4}\right)\varepsilon
$$

$$
+ \left(\frac{7}{2}n_f + 3a_0^2 - \frac{1}{2}\zeta_2 a_0^2 + \frac{5}{8}\right)\varepsilon^2 + \cdots
$$

The situation here is simpler than in Sect. 4.3: the terms with g_1^2 and with g_2 in $\varepsilon^2 f_2(\varepsilon)$ are separately finite at $\varepsilon \to 0$, so that we can put $g_2 = g_1^2$ with the ε^2 accuracy, and ζ_3 does not appear.

The propagator expressed via the renormalized quantities $\alpha(\mu)$, $a(\mu)$ is, with the α^2 accuracy,

$$
\not{p} S(p) = 1 + \frac{\alpha(\mu)}{4\pi\varepsilon} e^{-L\varepsilon} Z_\alpha \varepsilon e^{\gamma\varepsilon} \left[f_1'(\varepsilon) + f_1''(\varepsilon) Z_A a(\mu)\right]
$$

$$
+ \left(\frac{\alpha}{4\pi\varepsilon}\right)^2 e^{-2L\varepsilon} \varepsilon^2 e^{2\gamma\varepsilon} f_2(\varepsilon) \,,
$$

see (4.20). At $L = 0$ we have

$$\not{p}S(p) = 1$$
$$+ \frac{\alpha(\mu)}{4\pi\varepsilon} \left[c'_{10} + c''_{10}a(\mu) + (c'_{11} + c''_{11}a(\mu))\varepsilon + (c'_{12} + c''_{12}a(\mu))\varepsilon^2 + \cdots \right]$$
$$+ \left(\frac{\alpha}{4\pi\varepsilon} \right)^2 \left[c_{20} + c_{21}\varepsilon + c_{22}\varepsilon^2 + \cdots - \beta_0 \left(c_{10} + c_{11}\varepsilon + c_{12}\varepsilon^2 + \cdots \right) \right.$$
$$\left. + \frac{1}{2}\gamma_{A0}a \left(c''_{10} + c''_{11}\varepsilon + c''_{12}\varepsilon^2 + \cdots \right) \right].$$

Here $c_{1n} = c'_{1n} + c''_{1n}a(\mu)$; in c_{2n}, we may substitute $a(\mu)$ instead of a_0. This should be equal to $Z_\psi(\alpha(\mu), a(\mu))\not{p}S_r(p; \mu)$. Equating α terms, we obtain (4.23); equating α^2 terms, we now have

$$z_{20} = c_{20} - \beta_0 c_{10} - \frac{1}{2}\gamma_{A0}c''_{10}a,$$

$$z_{21} = c_{21} - (c_{10} + \beta_0)c_{11} - \frac{1}{2}\gamma_{A0}c''_{11}a,$$

$$r_2 = c_{22} - (c_{10} + \beta_0)c_{12} - \frac{1}{2}\gamma_{A0}c''_{12}a.$$

There are extra terms as compared to the gauge-invariant case (4.24). In QED $\gamma_{A0} = -2\beta_0$.

We obtain

$$Z_\psi(\alpha, a) = 1 - a\frac{\alpha}{4\pi\varepsilon} + \left[\frac{a^2}{2} + \left(n_f + \frac{3}{4} \right)\varepsilon \right] \left(\frac{\alpha}{4\pi\varepsilon} \right)^2 + \cdots \qquad (4.52)$$

It is also easy to write down $\not{p}S_r(p; \mu)$ at $L = 0$ (i.e., $\mu^2 = -p^2$), or even at arbitrary μ, but we shall not do this here.

The anomalous dimension of the electron field is

$$\gamma_\psi = \frac{d\log Z_\psi}{d\log \mu} = \left[-2(\varepsilon + \beta(\alpha))\frac{\partial}{\partial\log\alpha} - \gamma_A(\alpha)\frac{\partial}{\partial\log a} \right]$$
$$\left[(z'_1 + z''_1 a)\frac{\alpha}{4\pi\varepsilon} + \left(z_{20} - \frac{1}{2}z_1^2 + z_{21}\varepsilon \right)\left(\frac{\alpha}{4\pi\varepsilon} \right)^2 \right]$$
$$= -2z_1\frac{\alpha}{4\pi} - 4\left[\left(z_{20} - \frac{1}{2}z_1^2 + \frac{1}{2}\beta_0 z_1 + \frac{1}{4}\gamma_{A0}z''_1 a \right)\frac{1}{\varepsilon} + z_{21} \right]\left(\frac{\alpha}{4\pi} \right)^2,$$

where $z_1 = z'_1 + z''_1 a$. It must be finite at $\varepsilon \to 0$; therefore, the two-loop term must satisfy the self-consistency relation

$$z_{20} = \frac{1}{2}z_1(z_1 - \beta_0) - \frac{1}{4}\gamma_{A0}z''_1 a, \qquad (4.53)$$

or

$$c_{20} = \frac{1}{2}c_{10}(c_{10} + \beta_0) + \frac{1}{4}\gamma_{A0}c''_{10}a. \tag{4.54}$$

The renormalization constant of a non-gauge-invariant quantity (such as the electron field) must have the form, up to two loops,

$$
\begin{aligned}
Z_\psi = 1 &- \frac{1}{2}\gamma_{\psi 0}\frac{\alpha}{4\pi\varepsilon} \\
&+ \frac{1}{8}\left[\gamma_{\psi 0}(\gamma_{\psi 0} + 2\beta_0) + \gamma_{A0}\gamma''_{\psi 0}a - 2\gamma_{\psi 1}\varepsilon\right]\left(\frac{\alpha}{4\pi\varepsilon}\right)^2 + \cdots
\end{aligned} \tag{4.55}
$$

where $\gamma_{\psi 0} = \gamma'_{\psi 0} + \gamma''_{\psi 0}a$ (this formula generalizes the gauge-invariant case (4.30)).

We see that the self-consistency condition is indeed satisfied, and

$$\gamma_\psi(\alpha, a) = 2a\frac{\alpha}{4\pi} - (4n_f + 3)\left(\frac{\alpha}{4\pi}\right)^2 + \cdots \tag{4.56}$$

In the normal QED, with just one charged lepton field,

$$\gamma_\psi(\alpha, a) = 2a\frac{\alpha}{4\pi} - 7\left(\frac{\alpha}{4\pi}\right)^2 + \cdots \tag{4.57}$$

The electron mass anomalous dimension can be found in a similar way. We can calculate $\Sigma_S(p^2)$ at two loops (neglecting m^2) by retaining m_0 in the numerator of a single electron propagator in Fig. 4.13 and setting $m_0 \to 0$ in all the other places. This single m_0 has to be somewhere along the electron line which goes through all the diagrams, not in the electron loop in the first diagram: we need one helicity flip of the external electron, and one helicity flip in a loop yields zero contribution. Then we extract Z_m from (2.86). It must be gauge-invariant, because $m(\mu)$ is gauge-invariant. The result is

$$\gamma_m(\alpha) = 6\frac{\alpha}{4\pi} - \left(\frac{20}{3}n_f - 3\right)\left(\frac{\alpha}{4\pi}\right)^2 + \cdots \tag{4.58}$$

In the normal QED, with just one charged lepton field,

$$\gamma_m(\alpha) = 6\frac{\alpha}{4\pi} - \frac{11}{3}\left(\frac{\alpha}{4\pi}\right)^2 + \cdots \tag{4.59}$$

4.7 Two-loop corrections in QCD

The one-loop β-function in QCD is (3.45); the two-loop one is

$$\beta_1 = \frac{34}{3}C_A^2 - 4C_F T_F n_f - \frac{20}{3}C_A T_F n_f. \tag{4.60}$$

The second term follows from the QED result (4.39); non-abelian terms (containing C_A) are more difficult to derive.

How to express $\alpha_s(\mu')$ via $\alpha_s(\mu)$, if the scales μ and μ' are not too widely separated? We need to solve the RG equation (3.46) with the initial condition at μ. Let's introduce short notation:

$$a_s = \frac{\alpha_s(\mu)}{4\pi}, \quad a_s' = \frac{\alpha_s(\mu')}{4\pi}, \quad b_1 = \frac{\beta_1}{\beta_0}.$$

Then the integral of our equation is

$$-\int_{a_s}^{a_s'} \frac{1}{1 + b_1 a_s + \cdots} \frac{da_s}{a_s^2} = 2\beta_0 \log \frac{\mu'}{\mu}. \tag{4.61}$$

It is natural to introduce notation

$$l = 2\beta_0 \log \frac{\mu'}{\mu}.$$

Expanding the left-hand side, we obtain

$$-\int_{a_s}^{a_s'} (1 - b_1 a_s + \cdots) \frac{da_s}{a_s^2} = \frac{1}{a_s'} - \frac{1}{a_s} + b_1 \log \frac{a_s'}{a_s} + \cdots = l. \tag{4.62}$$

The solution is a series:

$$a_s' = a_s \left(1 + c_1 a_s + c_2 a_s^2 + \cdots\right).$$

Substituting it into (4.62), we obtain

$$\frac{1}{a_s}\left[1 - c_1 a_s + (c_1^2 - c_2)a_s^2 + \cdots\right] - \frac{1}{a_s} + b_1 \left[c_1 a_s + \cdots\right] = l$$

Therefore, $c_1 = -l$ and $c_2 = l^2 - b_1 l$. The final result is

$$a_s' = a_s \left[1 - l a_s + (l^2 - b_1 l)a_s^2 + \cdots\right]. \tag{4.63}$$

It is easy to find $a_s^3, \ldots$ corrections; they will contain $b_2, \ldots$

It is not possible to find the exact solution of the RG equation for $a_s(\mu)$ in elementary functions, if two (or more) terms in $\beta(a_s)$ are kept[2].

---

[2]With two terms, the solution can be written via the Lambert W-function.

However, an implicit solution can be obtained. Separating variables in the RG equation (3.46), we can write it as

$$\frac{1}{2\beta(a_s)}\frac{da_s}{a_s} = -d\log\mu\,. \tag{4.64}$$

Let's subtract and add two first terms in the expansion of the integrand in a_s:

$$\int_0^{a_s(\mu)}\left(\frac{1}{2\beta(a_s)} - \frac{1}{2\beta_0 a_s} + \frac{\beta_1}{2\beta_0^2}\right)\frac{da_s}{a_s}$$
$$-\frac{1}{2\beta_0 a_s(\mu)} - \frac{\beta_1}{2\beta_0^2}\log\left[\beta_0 a_s(\mu)\right] = -\log\frac{\mu}{\Lambda_{\overline{\mathrm{MS}}}}\,. \tag{4.65}$$

The added terms are explicitly integrated; the difference behaves well at $a_a \to 0$, and can be integrated from 0. The integration constant $\Lambda_{\overline{\mathrm{MS}}}$ has appeared here, as in the one-loop case (3.47). We can solve for this constant:

$$\Lambda_{\overline{\mathrm{MS}}} = \mu\exp\left(-\frac{1}{2\beta_0 a_s(\mu)}\right)\left[\beta_0 a_s(\mu)\right]^{-\beta_1/(2\beta_0^2)} K(a_s(\mu))\,, \tag{4.66}$$

where

$$K(a_s) = \exp\int_0^{a_s}\left(\frac{1}{2\beta(a_s)} - \frac{1}{2\beta_0 a_s} + \frac{\beta_1}{2\beta_0^2}\right)\frac{da_s}{a_s} = 1 + \cdots \tag{4.67}$$

This is a regular series in a_s; its a_s term contains β_2, and so on.

The quark mass anomalous dimension is, up to two loops,

$$\gamma_m = 6C_F\frac{\alpha_s}{4\pi} + C_F\left(3C_F + \frac{97}{3}C_A - \frac{20}{3}T_F n_f\right)\left(\frac{\alpha_s}{4\pi}\right)^2 + \cdots \tag{4.68}$$

Everything here, except the C_A term in γ_{m1}, can be obtained from QED result (4.58) by inserting obvious colour factors. The solution (2.96) of the RG equation can be written in the following convenient form. Let's subtract and add the first term of the expansion of the integrand in α_s; the difference can be integrated from 0:

$$m(\mu') = m(\mu)\left(\frac{\alpha(\mu')}{\alpha(\mu)}\right)^{\gamma_{m0}/(2\beta_0)} K_{\gamma_m}(\alpha_s(\mu'))K_{\gamma_m}^{-1}(\alpha_s(\mu))\,, \tag{4.69}$$

where for any anomalous dimension

$$\gamma(\alpha_s) = \gamma_0\frac{\alpha_s}{4\pi} + \gamma_1\left(\frac{\alpha_s}{4\pi}\right)^2 + \cdots$$

we define

$$K_\gamma(\alpha_s) = \exp \int_0^{\alpha_s} \left(\frac{\gamma(\alpha_s)}{2\beta(\alpha_s)} - \frac{\gamma_0}{2\beta_0} \right) \frac{d\alpha_s}{\alpha_s}$$
$$= 1 + \frac{\gamma_0}{2\beta_0} \left(\frac{\gamma_1}{\gamma_0} - \frac{\beta_1}{\beta_0} \right) \frac{\alpha_s}{4\pi} + \cdots \qquad (4.70)$$

This function has the obvious properties

$$K_0(\alpha_s) = 1, \quad K_{-\gamma}(\alpha_s) = K_\gamma^{-1}(\alpha_s), \quad K_{\gamma_1+\gamma_2}(\alpha_s) = K_{\gamma_1}(\alpha_s)K_{\gamma_2}(\alpha_s).$$

The solution (4.69) can also be rewritten as

$$m(\mu) = \hat{m} \left(\frac{\alpha_s(\mu)}{4\pi} \right)^{\gamma_{m0}/(2\beta_0)} K_{\gamma_m}(\alpha_s(\mu)), \qquad (4.71)$$

where $\hat{m}$ is a renormalization group invariant which characterizes the given quark flavour. The running mass $m(\mu)$ decreases when μ increases, because $\gamma_{m0} = 6C_F > 0$, $\beta_0 > 0$, so that the exponent in (4.71) is positive, and $\alpha_s(\mu)$ decreases with μ.

The QCD β- and γ-functions are also known at three [Tarasov *et al.* (1980); Tarasov (1982); Larin and Vermaseren (1993)] and four [van Ritbergen *et al.* (1997); Chetyrkin (1997); Vermaseren *et al.* (1997); Chetyrkin and Retey (2000); Chetyrkin (2005); Czakon (2005)] loops.

Chapter 5

On-shell renormalization scheme

5.1 On-shell renormalization of photon field

We shall consider QED with non-zero electron mass (2.81). Until now, we used $\overline{\text{MS}}$ renormalization scheme. In this scheme, the electron mass $m(\mu)$ and the coupling $\alpha(\mu)$ depend on the renormalization scale μ. It would be more exact to call $\overline{\text{MS}}$ a one-parameter family of renormalization schemes. But what are the experimentally measured electron mass $m \approx 0.511\,\text{MeV}$ and coupling $\alpha \approx 1/137$? They are defined in another renormalization scheme — the on-shell scheme. It contains no parameters. $\overline{\text{MS}}$ scheme is most useful at high energies, when the electron mass can be neglected (or considered as a small correction); μ should be of order of characteristic energies. When energies are of order m, the on-shell scheme is often more convenient. We shall discuss it now.

The photon field renormalized in the on-shell scheme A_{os} is related to the bare one A_0 by

$$A_0 = (Z_A^{\text{os}})^{1/2} A_{\text{os}} , \qquad (5.1)$$

where Z_A^{os} is the renormalization constant (it is not minimal in the sense (2.8)). The renormalized propagator is related to the bare one by

$$D_\perp(p^2) = Z_A^{\text{os}} D_\perp^{\text{os}}(p^2) . \qquad (5.2)$$

The bare photon propagator near the mass shell is

$$D_\perp(p^2) = \frac{1}{1 - \Pi(p^2)} \frac{1}{p^2} = \frac{1}{1 - \Pi(0)} \frac{1}{p^2} + \cdots . \qquad (5.3)$$

We require, by definition, that $D_\perp^{\text{os}}(p^2)$ behaves as the free propagator $1/p^2$ near the mass shell. Therefore, the photon-field renormalization constant in the on-shell scheme is

$$Z_A^{\text{os}} = \frac{1}{1 - \Pi(0)}.$$ (5.4)

How can we calculate $\Pi(0)$? From (2.15) we have

$$\Pi_\mu^\mu(p) = (d - 1)p^2\Pi(p^2),$$

and hence

$$\frac{\partial}{\partial p_\nu} \frac{\partial}{\partial p^\nu} \Pi_\mu^\mu(p)\bigg|_{p=0} = 2d(d - 1)\Pi(0).$$ (5.5)

The left-hand side has zero external momenta — it is a vacuum diagram, with the only energy scale m. It is not difficult to calculate a few more terms of expansion of $\Pi(p^2)$ in p^2 in the same manner.

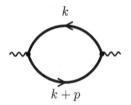

Fig. 5.1   One-loop photon self-energy

At one loop (Fig. 5.1), setting $m_0 = 1$ (it will be restored by dimensionality), we have

$$\Pi(0) = \frac{ie_0^2}{2d(d - 1)} \int \frac{d^d k}{(2\pi)^d} \operatorname{Tr} \gamma_\mu S_0(k) \gamma^\mu \left[\frac{\partial}{\partial p_\nu} \frac{\partial}{\partial p^\nu} S_0(k + p)\right]_{p=0}.$$ (5.6)

Expanding the free electron propagator $S_0(k + p)$ (2.82) in p,

$$S_0(k + p) = \frac{\not k + 1}{k^2 - 1} + \frac{\not p}{k^2 - 1} - \frac{2p \cdot k (\not k + 1)}{(k^2 - 1)^2}$$
$$- \frac{2p \cdot k \not p}{(k^2 - 1)^2} - \frac{p^2(\not k + 1)}{(k^2 - 1)^2} + \frac{(2p \cdot k)^2(\not k + 1)}{(k^2 - 1)^3} + \mathcal{O}(p^3),$$

we obtain

$$\left[\frac{\partial}{\partial p_\nu} \frac{\partial}{\partial p^\nu} S_0(k + p)\right]_{p=0} = -\frac{2d(\not k + 1) + 4\not k}{(k^2 - 1)^2} + \frac{8k^2(\not k + 1)}{(k^2 - 1)^3}.$$

Using also (1.55), (1.54) to simplify $\gamma_\mu S_0(k)\gamma^\mu$, we can easily calculate the trace:

$$
\begin{aligned}
\Pi(0) &= \frac{4ie_0^2}{d(d-1)} \int \frac{d^dk}{(2\pi)^d} \left[\frac{8}{D^4} + \frac{4(d-3)}{D^3} + \frac{d^2-4d+4}{D^2} \right] \\
&= -4 \frac{e_0^2}{(4\pi)^{d/2}} \left[8V(4) + 4(d-3)V(3) + (d^2-4d+4)V(2) \right] ,
\end{aligned}
$$

where $D = 1 - k^2$, and the definition (1.2) of the vacuum integrals was used. Using (1.9), we can express all $V(n)$ via $V(2) = \Gamma(\varepsilon)$; also restoring the power of m_0 by dimensionality, we finally arrive at

$$
\Pi(0) = -\frac{4}{3} \frac{e_0^2 m_0^{-2\varepsilon}}{(4\pi)^{d/2}} \Gamma(\varepsilon) . \tag{5.7}
$$

The ultraviolet divergence of the diagram of Fig. 2.4 does not depend on masses and external momenta, and is the same in (5.7) as in (2.28).

Therefore, the photon-field renormalization constant in the on-shell scheme is, with one-loop accuracy,

$$
Z_A^{\text{os}} = 1 - \frac{4}{3} \frac{e_0^2 m_0^{-2\varepsilon}}{(4\pi)^{d/2}} \Gamma(\varepsilon) . \tag{5.8}
$$

The renormalized photon propagator in the $\overline{\text{MS}}$ scheme and in the on-shell scheme are both ultraviolet-finite, therefore, the ratio $Z_A^{\text{os}}/Z_A(\alpha(\mu))$ must be finite at $\varepsilon \to 0$ if both of them are expressed via renormalized quantities (the on-shell photon propagator has no IR divergences because the photon is neutral). Therefore, we can reproduce the $\overline{\text{MS}}$ renormalization constant (2.32) from (5.8).

5.2 One-loop massive on-shell propagator diagram

Before discussing the electron mass and field renormalization in the on-shell scheme, we have to learn how to calculate the relevant diagrams.

Let's consider the on-shell propagator integral (Fig. 5.2, $p^2 = m^2$)

$$
\begin{aligned}
\int \frac{d^dk}{D_1^{n_1} D_2^{n_2}} &= i\pi^{d/2} m^{d-2(n_1+n_2)} M(n_1, n_2) , \\
D_1 &= m^2 - (k+p)^2 = -k^2 - 2p\cdot k , \quad D_2 = -k^2 .
\end{aligned} \tag{5.9}
$$

The power of m is evident from the dimensional counting, and our aim is to find the dimensionless function $M(n_1, n_2)$; we can put $m = 1$ to simplify

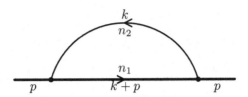

Fig. 5.2   One-loop on-shell propagator diagram

the calculation. It vanishes if n_1 is a non-positive integer.

Using Wick rotation and α-parametrization (1.5), we rewrite the definition (5.9) as

$$M(n_1, n_2) = \frac{\pi^{-d/2}}{\Gamma(n_1)\Gamma(n_2)} \int e^{-\alpha_1(k^2 + 2p \cdot k) - \alpha_2 k^2} \alpha_1^{n_1-1} \alpha_2^{n_2-1} d\alpha_1 \, d\alpha_2 \, d^d k \,.$$

(5.10)

We want to separate a full square in the exponent; to this end, we shift the integration momentum:

$$k' = k + \frac{\alpha_1}{\alpha_1 + \alpha_2} p$$

and obtain

$$M(n_1, n_2)$$
$$= \frac{\pi^{-d/2}}{\Gamma(n_1)\Gamma(n_2)} \int \exp\left[-\frac{\alpha_1^2}{\alpha_1 + \alpha_2}\right] \alpha_1^{n_1-1} \alpha_2^{n_2-1} d\alpha_1 \, d\alpha_2$$
$$\times \int e^{-(\alpha_1 + \alpha_2) k'^2} d^d k'$$
$$= \frac{1}{\Gamma(n_1)\Gamma(n_2)} \int \exp\left[-\frac{\alpha_1^2}{\alpha_1 + \alpha_2}\right] \frac{\alpha_1^{n_1-1} \alpha_2^{n_2-1}}{(\alpha_1 + \alpha_2)^{d/2}} d\alpha_1 \, d\alpha_2 \,.$$

Substituting $\eta = \alpha_1 + \alpha_2$, $\alpha_1 = \eta x$, $\alpha_2 = \eta(1 - x)$, we get

$$M(n_1, n_2) = \frac{1}{\Gamma(n_1)\Gamma(n_2)} \int_0^1 x^{n_1-1}(1-x)^{n_2-1} dx$$
$$\times \int_0^\infty e^{-\eta x^2} \eta^{-d/2 + n_1 + n_2 - 1} d\eta$$
$$= \frac{\Gamma(-d/2 + n_1 + n_2)}{\Gamma(n_1)\Gamma(n_2)} \int_0^1 x^{d - n_1 - 2n_2 - 1}(1 - x)^{n_2-1} dx \,.$$

Finally, we arrive at

$$M(n_1, n_2) = \frac{\Gamma(-d/2 + n_1 + n_2)\Gamma(d - n_1 - 2n_2)}{\Gamma(n_1)\Gamma(d - n_1 - n_2)}. \tag{5.11}$$

The denominator in (5.9) behaves as $(k^2)^{n_1+n_2}$ at $k \to \infty$. Therefore, the integral diverges if $d \geq 2(n_1 + n_2)$. At $d \to 4$ this means $n_1 + n_2 \leq 2$. This ultraviolet divergence shows itself as a $1/\varepsilon$ pole of the first Γ function in the numerator of (5.11) (this Γ function depends on $n_1 + n_2$, i.e., on the behaviour of the integrand at $k \to \infty$). The integral (5.9) can also have infrared divergences. Its denominator behaves as $k^{n_1+2n_2}$ at $k \to 0$, and the integral diverges in this region if $d \leq n_1 + 2n_2$. At $d \to 4$ this means $n_1 + 2n_2 \geq 4$. This infrared divergence shows itself as a $1/\varepsilon$ pole of the second Γ function in the numerator of (5.11) (this Γ function depends on $n_1 + 2n_2$, i.e., on the behaviour of the integrand at $k \to 0$).

Fig. 5.3 One-loop on-shell propagator diagram

Fig. 5.4 The basis integral

Let's summarize. There is one generic topology of one-loop massive on-shell propagator diagrams in QED and QCD (Fig. 5.3). All Feynman integrals of this class, with any integer indices n_1, n_2, are proportional to V_1 (1.10) (Fig. 5.4), with coefficients being rational functions of d. For example, for $M(1,1)$,

$$= -\frac{1}{2}\frac{d - 2}{d - 3} \tag{5.12}$$

Two-loop [Gray *et al.* (1990); Broadhurst *et al.* (1991); Broadhurst (1992)] and three-loop [Laporta and Remiddi (1996); Melnikov and van

Ritbergen (2000)] massive on-shell diagrams can be calculated using integration by parts.

5.3   On-shell renormalization of electron mass and field

The electron mass m in the on-shell renormalization scheme is defined as the position of the pole of the electron propagator. On-shell external electron lines have $p^2 = m^2$; it is convenient to use the free propagator containing the same quantity m rather than m_0. Therefore, we rewrite the Lagrangian (2.81) as

$$L = \bar{\psi}_0(i\not{D} - m)\psi_0 + \delta m \bar{\psi}_0 \psi_0 , \quad \delta m = m - m_0 , \qquad (5.13)$$

and consider the mass counterterm not as a part of the unperturbed Lagrangian, but as a perturbation. Then the free electron propagator

$$S_0(p) = \frac{1}{\not{p} - m} = \frac{\not{p} + m}{p^2 - m^2} \qquad (5.14)$$

contains m, and the mass counterterm produces the vertex

$= i\,\delta m . \qquad (5.15)$

The mass renormalization constant in the on-shell scheme is defined as usual:

$$m_0 = Z_m^{\mathrm{os}} m . \qquad (5.16)$$

It is more convenient to write the electron self-energy (2.84) in the form

$$\Sigma(p) = m\Sigma_1(p^2) + (\not{p} - m)\Sigma_2(p^2) \qquad (5.17)$$

now. The electron propagator

$$S(p) = \frac{1}{[1 - \Sigma_2(p^2)](\not{p} - m) + \delta m - m\Sigma_1(p^2)} \qquad (5.18)$$

has a pole at $p^2 = m^2$ if

$$\delta m = m\Sigma_1(m^2) , \qquad (5.19)$$

or

$$Z_m^{\mathrm{os}} = 1 - \frac{\delta m}{m} = 1 - \Sigma_1(m^2) . \qquad (5.20)$$

The equation (5.19) can be solved for δm by iterations (at higher orders, its right-hand side contains δm because of the vertex (5.15)).

Near the mass shell, we can expand $\Sigma_1(p^2)$ as

$$\Sigma_1(p^2) - \frac{\delta m}{m} = \Sigma_1'(m^2)(p^2 - m^2) + \cdots \tag{5.21}$$

so that

$$S(p) = \frac{1}{1 - \Sigma_2(m^2) - 2m^2\Sigma_1'(m^2)} \frac{\not{p} + m}{p^2 - m^2} + \cdots \tag{5.22}$$

The electron field renormalized in the on-shell scheme,

$$\psi_0 = \left(Z_\psi^{\text{os}}\right)^{1/2} \psi_{\text{os}}, \quad S(p) = Z_\psi^{\text{os}} S_{\text{os}}(p), \tag{5.23}$$

is defined in such a way that its propagator $S_{\text{os}}(p)$ behaves as $S_0(p)$ (5.14) near the mass shell:

$$Z_\psi^{\text{os}} = \left[1 - \Sigma_2(m^2) - 2m^2\Sigma_1'(m^2)\right]^{-1}. \tag{5.24}$$

In order to calculate Z_m^{os} and Z_ψ^{os}, it is convenient to introduce the function

$$\begin{aligned} T(t) &= \frac{1}{4m} \operatorname{Tr}(\not{p} + 1)\Sigma(mv(1 + t)) \\ &= \Sigma_1(m^2) + \left[\Sigma_2(m^2) + 2m^2\Sigma_1'(m^2)\right] t + \cdots \end{aligned} \tag{5.25}$$

so that

$$Z_m^{\text{os}} = 1 - T(0), \quad Z_\psi^{\text{os}} = \left[1 - T'(0)\right]^{-1}. \tag{5.26}$$

Let's calculate it at one loop (Fig. 5.5). We put $m = 1$; the power of m will be restored by dimensionality:

$$T(t) = -ie_0^2 \int \frac{d^d k}{(2\pi)^d} \frac{1}{D_1(t)\, D_2} \frac{1}{4} \operatorname{Tr}(\not{p}+1)\gamma^\mu(\not{k}+\not{p}+1)\gamma^\nu \left(g_{\mu\nu} + \xi\frac{k_\mu k_\nu}{D_2}\right), \tag{5.27}$$

where

$$p = v(1 + t), \quad D_1(t) = 1 - (k + p)^2, \quad D_2 = -k^2.$$

In calculating the numerator, we can express

$$p \cdot k = \frac{1}{2}\left[D_2 - D_1(t) + 1 - (1 + t)^2\right]$$

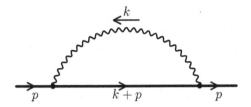

Fig. 5.5   One-loop electron self-energy

and omit terms with $D_1(t)$, because the resulting integrals contain no scale. We obtain

$$T(t) = -ie_0^2 \int \frac{d^d k}{(2\pi)^d} \frac{1}{D_1(t)} \left[\frac{2}{D_2} - \frac{d-2}{2}(1-t) + \mathcal{O}(t^2) \right].$$

This result is gauge-independent.

Expanding

$$D_1(t) = D_1 + (D_1 - D_2 - 2)t + \mathcal{O}(t^2), \quad D_1 = 1 - (k+v)^2,$$

we obtain

$$T(t) = -ie_0^2 \int \frac{d^d k}{(2\pi)^d} \left[\frac{2(1-t)}{D_1 D_2} - \frac{(d-2)(1-2t)}{2D_1} + \frac{4t}{D_1^2 D_2} \right.$$
$$\left. - \frac{(d-4)t}{D_1^2} - \frac{(d-2)D_2 t}{2D_1^2} + \mathcal{O}(t^2) \right].$$

Using (5.11) and restoring the power of m, we finally arrive at

$$T(t) = \frac{e_0^2 m^{-2\varepsilon}}{(4\pi)^{d/2}} \Gamma(\varepsilon) \frac{d-1}{d-3} (1-t) + \mathcal{O}(t^2). \tag{5.28}$$

Therefore,

$$Z_m^{\mathrm{os}} = Z_\psi^{\mathrm{os}} = 1 - \frac{e_0^2 m^{-2\varepsilon}}{(4\pi)^{d/2}} \Gamma(\varepsilon) \frac{d-1}{d-3}. \tag{5.29}$$

The fact that $Z_m^{\mathrm{os}} = Z_\psi^{\mathrm{os}}$ at one loop is a pure accident; at two loops it is no longer so [Gray *et al.* (1990); Broadhurst *et al.* (1991)]. The on-shell mass m is a measurable quantity, and hence gauge-invariant; therefore, Z_m^{os} is gauge-invariant to all orders. It has been proved that Z_ψ^{os} is also gauge-invariant in QED (but not in QCD).

What's the relation between the on-shell mass m and the $\overline{\text{MS}}$ mass $m(\mu)$?

$$m_0 = Z_m(\alpha(\mu))m(\mu) = Z_m^{\text{os}}m\,,$$

and therefore

$$m(\mu) = \frac{Z_m^{\text{os}}}{Z_m(\alpha(\mu))}m\,. \tag{5.30}$$

We have to re-express Z_m^{os} via $\alpha(\mu)$; then the ratio of renormalization constants is finite at $\varepsilon \to 0$, because both renormalized masses are finite. At one loop

$$\frac{m(\mu)}{m} = 1 - 6\frac{\alpha}{4\pi}\left(\log\frac{\mu}{m} + \frac{2}{3}\right)\,. \tag{5.31}$$

This formula is OK when $\mu \sim m$; otherwise, it is much better to relate $m(m)$ to m, and then to solve the RG equation with the initial condition $m(m)$.

5.4 On-shell charge

On the mass shell ($p^2 = m^2$), $\Gamma^\mu(p,p)$ has only one γ-matrix structure,

$$\Gamma^\mu(p,p) = Z_\Gamma^{\text{os}}\gamma^\mu\,, \tag{5.32}$$

if sandwiched between $\bar{u}_2 \ldots u_1$ which satisfy the Dirac equation (see Sect. 5.5 for more details). The physical matrix element of scattering of an electron by a photon is

$$e_0\Gamma^\mu Z_\psi^{\text{os}}\left(Z_A^{\text{os}}\right)^{1/2}\,.$$

The only case when all 3 particles are on-shell is $p^2 = m^2$, and the photon momentum $q \to 0$. By definition, this matrix element is

$$e\gamma^\mu\,, \tag{5.33}$$

where e is the renormalized charge in the on-shell scheme. It is related to the bare charge by

$$e_0 = \left(Z_\alpha^{\text{os}}\right)^{1/2}e\,. \tag{5.34}$$

Therefore,

$$Z_\alpha^{\text{os}} = \left(Z_\Gamma^{\text{os}}Z_\psi^{\text{os}}\right)^{-2}\left(Z_A^{\text{os}}\right)^{-1}\,. \tag{5.35}$$

On-shell electron charge is measured in macroscopic experiments with smooth electromagnetic fields (having $q \to 0$); it is

$$\alpha = \frac{e^2}{4\pi} \approx \frac{1}{137} \,. \tag{5.36}$$

In QED, the situation is simplified by the Ward identity (2.64). Near the mass shell

$$S(p) = \frac{Z_\psi^{\mathrm{os}}}{\not{p} - m} \,,$$

and hence

$$\Gamma^\mu(p,p) = \left(Z_\psi^{\mathrm{os}}\right)^{-1} \gamma^\mu \,.$$

Therefore,

$$Z_\psi^{\mathrm{os}} Z_\Gamma^{\mathrm{os}} = 1 \,, \tag{5.37}$$

and

$$Z_\alpha^{\mathrm{os}} = (Z_A^{\mathrm{os}})^{-1} \,. \tag{5.38}$$

At one loop, from (5.8),

$$Z_\alpha^{\mathrm{os}} = 1 + \frac{4}{3} \frac{e_0^2 m_0^{-2\varepsilon}}{(4\pi)^{d/2}} \Gamma(\varepsilon) \,. \tag{5.39}$$

The $\overline{\mathrm{MS}}$ (running) coupling $\alpha(\mu)$ is related to the on-shell coupling α by

$$\alpha(\mu) = \alpha \frac{Z_\alpha^{\mathrm{os}}}{Z_\alpha(\mu)} = \alpha \left[1 + \frac{4}{3} \frac{\alpha(\mu)}{4\pi} \left(\left(\frac{\mu}{m}\right)^{2\varepsilon} e^{\gamma\varepsilon} \Gamma(\varepsilon) - \frac{1}{\varepsilon} \right) \right] \,.$$

Therefore

$$\alpha(\mu) = \alpha \left[1 + \frac{8}{3} \frac{\alpha}{4\pi} \log \frac{\mu}{m} \right] \,. \tag{5.40}$$

We can always find the μ-dependence from the RG equation, the initial condition is $\alpha(m) = \alpha$ at one loop.

5.5 Magnetic moment

Let's consider scattering of an on-shell electron in electromagnetic field. The physical scattering amplitude is

$$e_0 Z_\psi^{\text{os}} (Z_A^{\text{os}})^{1/2} \, \bar{u}' \Gamma^\mu(p, p') u \,, \tag{5.41}$$

where $p^2 = m^2$, $p'^2 = m^2$, and the initial and final electron wave functions satisfy the Dirac equation

$$\not{p} u = m u \,, \quad \bar{u}' \not{p}' = \bar{u}' m \,. \tag{5.42}$$

We can substitute the on-shell charge

$$e = e_0 \, (Z_A^{\text{os}})^{1/2} \,.$$

Using the Ward identity (2.63) and the electron self-energy (5.17), we obtain

$$\Gamma^\mu(p, p') q_\mu = (\not{p}' - \not{p})(1 - \Sigma_2(m^2)) \,,$$

where $q = p' - p$, and hence for on-shell electron wave functions (5.42)

$$\bar{u}' \Gamma^\mu q_\mu u = 0 \,. \tag{5.43}$$

What γ-matrix structures can $\bar{u}' \Gamma^\mu u$ have? Using the Dirac equations (5.42), we can always eliminate $\not{p}$ by anticommuting it to u, and $\not{p}'$ by anticommuting it to $\bar{u}'$. For example,

$$\bar{u}' \sigma^{\mu\nu} q_\nu u = i \, \bar{u}' \left((p + p')^\mu - 2m\gamma^\mu \right) u \,. \tag{5.44}$$

We are left with 3 structures: γ^μ, $(p + p')^\mu$, and $q^\mu = (p' - p)^\mu$. The last one is excluded by (5.43). Therefore,

$$Z_\psi^{\text{os}} \, \bar{u}' \Gamma^\mu u = \bar{u}' \left[(F_1(q^2) + F_2(q^2))\gamma^\mu - F_2(q^2) \frac{(p + p')^\mu}{2m} \right] u \,. \tag{5.45}$$

Using (5.44), we can also rewrite this as

$$\begin{aligned}
Z_\psi^{\text{os}} \, \bar{u}' \Gamma^\mu u &= \bar{u}' \left[F_1(q^2)\gamma^\mu + F_2(q^2) \frac{i\sigma^{\mu\nu} q_\nu}{2m} \right] u \\
&= \bar{u}' \left[F_1(q^2) \frac{(p + p')^\mu}{2m} + (F_1(q^2) + F_2(q^2)) \frac{i\sigma^{\mu\nu} q_\nu}{2m} \right] u \,.
\end{aligned} \tag{5.46}$$

Let's rewrite (5.45) as

$$\bar{u}' \Gamma^\mu u = \bar{u}' \left(\sum f_i T_i^\mu \right) u \,, \quad T_1^\mu = \frac{(p + p')^\mu}{2m} \,, \quad T_2^\mu = \gamma^\mu \,,$$

then

$$F_1(q^2) = Z_\psi^{\text{os}}(f_1 + f_2), \quad F_2(q^2) = -Z_\psi^{\text{os}} f_1. \tag{5.47}$$

If we introduce the traces

$$y_i = \frac{1}{4m^2} \operatorname{Tr} T'_{i\mu}(\not{p}' + m)\Gamma^\mu(\not{p} + m), \quad T'_{1\mu} = \frac{p_\mu}{m}, \quad T'_{2\mu} = \gamma_\mu,$$

then

$$y_i = M_{ij} f_j, \quad M_{ij} = \frac{1}{4m^2} \operatorname{Tr} T'_{i\mu}(\not{p}' + m)T_j^\mu(\not{p} + m),$$

and we can find f_i by solving the linear system:

$$f_i = \left(M^{-1}\right)_{ij} y_j.$$

Let's introduce the notation

$$t = -\frac{q^2}{4m^2}.$$

Calculating the traces, we find

$$M = 2 \begin{pmatrix} (1+t)^2 & 1+t \\ 1+t & 1-(d-2)t \end{pmatrix},$$

and hence

$$M^{-1} = -\frac{1}{2(d-2)t(1+t)^2} \begin{pmatrix} 1-(d-2)t & -(1+t) \\ -(1+t) & (1+t)^2 \end{pmatrix}.$$

Finally, we obtain from (5.47)

$$F_1(q^2) = \frac{Z_\psi^{\text{os}}}{2(d-2)(1+t)^2}$$
$$\times \frac{1}{4} \operatorname{Tr}\left[(d-1)v_\mu - (1+t)\gamma_\mu\right](\not{p}' + 1)\Gamma^\mu(\not{p} + 1), \tag{5.48}$$

$$F_2(q^2) = \frac{Z_\psi^{\text{os}}}{2(d-2)t(1+t)^2}$$
$$\times \frac{1}{4} \operatorname{Tr}\left[(1-(d-2)t)v_\mu - (1+t)\gamma_\mu\right](\not{p}' + 1)\Gamma^\mu(\not{p} + 1), \tag{5.49}$$

where $v = p/m$, $v' = p'/m$. At the tree level, $\Gamma^\mu = \gamma^\mu$, and, naturally, we obtain $F_1(q^2) = 1$, $F_2(q^2) = 0$. When calculating loop corrections, we can apply these projectors to integrands of vertex diagrams, and express $F_{1,2}(q^2)$ via scalar integrals.

The Dirac form factor (5.48) at $q^2 = 0$ is

$$F_1(0) = Z_\psi^{\text{os}} \frac{1}{4} \operatorname{Tr} v_\mu \Gamma^\mu(mv, mv)(\not{v} + 1).$$
(5.50)

Due to the Ward identity (2.64),

$$F_1(0) = Z_\psi^{\text{os}} \left[1 - v^\mu \frac{\partial}{\partial p^\mu} \frac{1}{4} \operatorname{Tr} \Sigma(p)(\not{v} + 1) \Big|_{p=mv} \right] = Z_\psi^{\text{os}} \left[1 - T'(0) \right] = 1,$$
(5.51)

see (5.25), (5.26). The total charge of electron is not changed by radiative corrections.

It may seem that the Pauli form factor (5.49) is singular at $q^2 \to 0$. Of course, it is not. Let's substitute the expansion

$$\Gamma^\mu(mv, mv + q) = \Gamma_0^\mu + \Gamma_1^{\mu\nu} \frac{q_\nu}{m} + \cdots$$

into (5.49), split $q = (q \cdot v)v + q_\perp$ (where $q \cdot v/m = 2t$, $q_\perp^2/m^2 = -4t(1+t)$), and average over the directions of $q_\perp$ in the $(d-1)$-dimensional subspace orthogonal to v:

$$\overline{\frac{q^\alpha}{m}} = 2tv^\alpha, \qquad \overline{\frac{q^\alpha q^\beta}{m^2}} = -\frac{4t}{d-1} \left[(1+t)g^{\alpha\beta} - (1+dt)v^\alpha v^\beta \right].$$

We obtain

$$F_2(0) = \frac{Z_\psi^{\text{os}}}{d-2} \left[\frac{1}{4} \operatorname{Tr}(\gamma_\mu - dv_\mu)\Gamma_0^\mu(\not{v} + 1) \right.$$
$$\left. + \frac{2}{d-1} \frac{1}{4} \operatorname{Tr} \left(\gamma_\mu \gamma_\nu + \gamma_\mu v_\nu - \gamma_\nu v_\mu - v_\mu v_\nu \right) \Gamma_1^{\mu\nu}(\not{v} + 1) \right].$$
(5.52)

This means that in order to calculate the anomalous magnetic moment we need the vertex and its first derivative in q at $q = 0$.

As already mentioned, the tree diagram ($\Gamma_0^\mu = \gamma^\mu$, $\Gamma_1^{\mu\nu} = 0$) does not contribute to $F_2(0)$. The first contributing diagram is one-loop (Fig. 5.6). All charged external lines are on-shell, therefore, this vertex diagram is gauge-invariant. We shall use Feynman gauge to simplify the calculation, and put $m = 1$ (the power of m will be restored by dimensionality). Then the one-loop vertex is

$$-ie_0^2 \int \frac{d^d k}{(2\pi)^d} \frac{1}{k^2} \gamma_\alpha S_0(k + v + q)\gamma^\mu S_0(k + v)\gamma^\alpha.$$

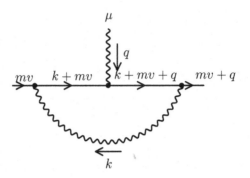

Fig. 5.6   One-loop anomalous magnetic moment

We need to expand it in q up to the linear term. In this diagram, only one propagator depends on q. Using

$$S_0(p+q) = S_0(p) + \frac{\slashed{q}}{p^2 - m^2} - \frac{\slashed{p} + m}{(p^2 - m^2)^2} \, 2p \cdot q + \cdots$$

for $p = k + mv$, we obtain $(m = 1)$

$$\Gamma_0^\mu = ie_0^2 \int \frac{d^d k}{(2\pi)^d} \frac{\gamma_\alpha(\slashed{k} + \slashed{v} + 1)\gamma^\mu(\slashed{k} + \slashed{v} + 1)\gamma^\alpha}{D_1^2 D_2},$$

$$\Gamma_1^{\mu\nu} = ie_0^2 \int \frac{d^d k}{(2\pi)^d} \frac{\gamma_\alpha \left[D_1 \gamma^\nu + 2(k+v)^\nu(\slashed{k} + \slashed{v} + 1) \right] \gamma^\mu(\slashed{k} + \slashed{v} + 1)\gamma^\alpha}{D_1^3 D_2},$$

where $D_1 = 1 - (k+v)^2$, $D_2 = -k^2$ (see (5.9)). We may replace $Z_\psi^{\text{os}} \to 1$ in (5.52), because corrections are beyond our accuracy; the two contributions to $F_2(0)$ are

$$\frac{ie_0^2}{d-2} \int \frac{d^d k}{(2\pi)^d} \frac{N_0}{D_1^2 D_2},$$

$$N_0 = \frac{1}{4} \text{Tr}(\gamma_\mu - dv_\mu)\gamma_\alpha(\slashed{k} + \slashed{v} + 1)\gamma^\mu(\slashed{k} + \slashed{v} + 1)\gamma^\alpha(\slashed{v} + 1),$$

and

$$\frac{2ie_0^2}{(d-1)(d-2)} \int \frac{d^d k}{(2\pi)^d} \frac{N_1}{D_1^3 D_2},$$

$$N_1 = \frac{1}{4} \text{Tr}(\gamma_\mu \gamma_\nu + \gamma_\mu v_\nu - \gamma_\nu v_\mu - v_\mu v_\nu)\gamma_\alpha$$
$$\times \left[D_1 \gamma^\nu + 2(k+v)^\nu(\slashed{k} + \slashed{v} + 1) \right] \gamma^\mu(\slashed{k} + \slashed{v} + 1)\gamma^\alpha(\slashed{v} + 1).$$

When calculating N_0, we may omit terms with D_1^2:

$$N_0 \Rightarrow -d(d-2)D_1D_2 - (d-1)(d-4)D_1 + \frac{1}{2}d(d-2)D_2^2 + d(d-3)D_2 - 4(d-1).$$

When calculating N_1, we may omit terms with D_1^3:

$$N_1 \Rightarrow \frac{1}{2}\Big[3(d-2)D_1^2D_2 + (d^3 - 8d^2 + 17d - 14)D_1^2 - 3(d-2)D_1D_2^2$$
$$- (d^3 - 8d^2 + 21d - 26)D_1D_2 + 4(d-1)^2D_1 + (d-2)D_2^3$$
$$+ 4(d-3)D_2^2 - 16D_2\Big].$$

Collecting all this together, we obtain

$$F_2(0) = -\frac{e_0^2}{(4\pi)^{d/2}}\frac{1}{d-2}\Big\{-d(d-1)M(1,0) - (d-1)(d-4)M(1,1)$$
$$+ \frac{1}{2}d(d-2)M(2,-1) + d(d-3)M(2,0) - 4(d-1)M(2,1)$$
$$+ \frac{1}{d-1}\Big[3(d-2)M(1,0) + (d^3 - 8d^2 + 17d - 14)M(1,1)$$
$$- 3(d-2)M(2,-1) - (d^3 - 8d^2 + 21d - 26)M(2,0)$$
$$+ 4(d-1)^2M(2,1) + (d-2)M(3,-2)$$
$$+ 4(d-3)M(3,-1) - 16M(3,0)\Big]\Big\}.$$

Using (5.11) and restoring the power of m, we finally arrive at the anomalous magnetic moment

$$F_2(0) = \frac{e_0^2 m^{-2\varepsilon}}{(4\pi)^{d/2}}\Gamma(\varepsilon)\frac{(d-4)(d-5)}{d-3} + \cdots \tag{5.53}$$

As expected, it is finite at $\varepsilon \to 0$:

$$F_2(0) = \frac{\alpha}{2\pi} + \cdots \tag{5.54}$$

As we can see from the second line in (5.46), the total magnetic moment of electron is

$$F_1(0) + F_2(0) = 1 + \frac{\alpha}{2\pi} + \cdots$$

(in units of Bohr magneton). It is not difficult to calculate the two-loop correction to the magnetic moment, if one has a program implementing the integration-by-parts evaluation of two-loop on-shell propagator integrals. The three-loop calculation [Laporta and Remiddi (1996)] was a major breakthrough.

5.6 Two-loop massive vacuum diagram

Let's consider the two-loop massive vacuum diagram (Fig. 5.7):

$$\int \frac{d^d k_1\, d^d k_2}{D_1^{n_1} D_2^{n_2} D_3^{n_3}} = -\pi^d m^{2(d-n_1-n_2-n_3)} V(n_1, n_2, n_3)\,,$$
$$D_1 = m^2 - k_1^2\,, \quad D_2 = m^2 - k_2^2\,, \quad D_3 = -(k_1 - k_2)^2\,. \tag{5.55}$$

The power of m is evident from the dimensional counting, and our aim is to find the dimensionless function $V(n_1, n_2, n_3)$; we can put $m = 1$ to simplify the calculation. It is symmetric with respect to $n_1 \leftrightarrow n_2$; it vanishes if n_1 or n_2 is a non-positive integer. Using Wick rotation and α-parametrization (1.5), we rewrite the definition (5.55) as

$$V(n_1, n_2, n_3) = \frac{\pi^{-d}}{\Gamma(n_1)\Gamma(n_2)\Gamma(n_3)} \int e^{-\alpha_1(k_1^2+1)-\alpha_2(k_2^2+1)-\alpha_3(k_1-k_2)^2}$$
$$\times \alpha_1^{n_1-1} \alpha_2^{n_2-1} \alpha_3^{n_3-1} d\alpha_1\, d\alpha_2\, d\alpha_3\, d^d k_1\, d^d k_2\,. \tag{5.56}$$

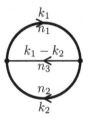

Fig. 5.7   Two-loop massive vacuum diagram

The integral in the loop momenta here has the form (1.23) with

$$A = \begin{pmatrix} \alpha_1 + \alpha_3 & -\alpha_3 \\ -\alpha_3 & \alpha_2 + \alpha_3 \end{pmatrix}\,,$$

and we obtain

$$V(n_1, n_2, n_3) = \frac{1}{\Gamma(n_1)\Gamma(n_2)\Gamma(n_3)} \int e^{-\alpha_1 - \alpha_2}$$

$$\times (\alpha_1\alpha_2 + \alpha_1\alpha_3 + \alpha_2\alpha_3)^{-d/2} \alpha_1^{n_1-1} \alpha_2^{n_2-1} \alpha_3^{n_3-1} d\alpha_1 \, d\alpha_2 \, d\alpha_3 .$$

To calculate this integral, it is most convenient to choose the "radial" variable $\eta = \alpha_1 + \alpha_2$ (Sect. 1.2), and substitute $\alpha_i = \eta x_i$. The integral in η is trivial:

$$V(n_1, n_2, n_3) = \frac{1}{\Gamma(n_1)\Gamma(n_2)\Gamma(n_3)} \int_0^\infty e^{-\eta} \eta^{-d+n_1+n_2+n_3-1} d\eta$$

$$\times \int (x_1 x_2 + x_1 x_3 + x_2 x_3)^{-d/2} x_1^{n_1-1} x_2^{n_2-1} x_3^{n_3-1} \delta(x_1 + x_2 - 1) dx_1 \, dx_2 \, dx_3$$

$$= \frac{\Gamma(-d + n_1 + n_2 + n_3)}{\Gamma(n_1)\Gamma(n_2)\Gamma(n_3)}$$

$$\times \int [x_1(1 - x_1) + x_3]^{-d/2} x_1^{n_1-1} (1 - x_1)^{n_2-1} x_3^{n_3-1} dx_1 \, dx_3 .$$

Substituting $x_3 = x_1(1 - x_1)y$, we get

$$V(n_1, n_2, n_3) = \frac{\Gamma(-d + n_1 + n_2 + n_3)}{\Gamma(n_1)\Gamma(n_2)\Gamma(n_3)} \int_0^\infty \frac{y^{n_3-1} dy}{(y + 1)^{d/2}}$$

$$\times \int_0^1 x^{-d/2+n_1+n_3-1} (1 - x)^{-d/2+n_2+n_3-1} dx$$

The integral

$$\int_0^\infty \frac{y^{n_3-1} dy}{(y + 1)^{d/2}} = \frac{\Gamma(n_3)\Gamma(d/2 - n_3)}{\Gamma(d/?)}$$

is easily calculated using the substitution $z = 1/(y + 1)$; the second integral is the Euler B-function. We arrive at the result [Vladimirov (1980)]

$$V(n_1, n_2, n_3) =$$

$$\frac{\Gamma\left(\frac{d}{2} - n_3\right) \Gamma\left(n_1 + n_3 - \frac{d}{2}\right) \Gamma\left(n_2 + n_3 - \frac{d}{2}\right) \Gamma(n_1 + n_2 + n_3 - d)}{\Gamma\left(\frac{d}{2}\right) \Gamma(n_1)\Gamma(n_2)\Gamma(n_1 + n_2 + 2n_3 - d)} . \quad (5.57)$$

This is the only class of two-loop diagrams for which a general formula for arbitrary n_i (not necessarily integer) is known.

Let's summarize. There is one generic topology of two-loop vacuum diagrams in QED and QCD (Fig. 5.8). All Feynman integrals of this class, with any integer indices n_i, are proportional to V_1^2 (1.10) (Fig. 5.9), with coefficients being rational functions of d.

Fig. 5.8   Two-loop vacuum diagram

Fig. 5.9   The basis integral

Three-loop massive vacuum diagrams can be calculated using integration by parts [Broadhurst (1992)].

5.7   On-shell renormalization of photon field and charge at two loops

Using the results of the previous Section, it is not (very) difficult to calculate the photon self-energy $\Pi(0)$ at two loops. We apply (5.5) to the diagrams of Fig. 4.10, calculate the derivatives in p, and reduce the problem to the vacuum integrals (5.55). The result is

$$\Pi_2(0) = -\frac{2}{3}\frac{e_0^4 m_0^{-4\varepsilon}}{(4\pi)^d}\Gamma^2(\varepsilon)\frac{(d-4)(5d^2-33d+34)}{d(d-5)}\,. \qquad (5.58)$$

Therefore, for the on-shell charge e we obtain

$$\frac{e_0^2}{e^2} = Z_\alpha^{\text{os}} = (Z_A^{\text{os}})^{-1} = 1 - \Pi(0)$$

$$= 1 + \frac{4}{3}\frac{e_0^2 m_0^{-2\varepsilon}}{(4\pi)^{d/2}}\Gamma(\varepsilon) - \frac{4}{3}\varepsilon\frac{9+7\varepsilon-10\varepsilon^2}{(2-\varepsilon)(1+2\varepsilon)}\left(\frac{e_0^2 m_0^{-2\varepsilon}}{(4\pi)^{d/2}}\Gamma(\varepsilon)\right)^2 + \cdots$$

$$(5.59)$$

where $\Pi(0)$ at one (5.7)) and two (5.58) was used. In the one-loop term, we have to substitute $e_0^2 = Z_\alpha^{\text{os}} e^2$ and $m_0 = Z_m^{\text{os}} m$, with one-loop Z_α^{os} (5.39) and Z_m^{os} (5.29). This results in

$$
\frac{e^2}{e_0^2} = (Z_\alpha^{\text{os}})^{-1}
$$

$$
= 1 - \frac{4}{3} \frac{e^2 m^{-2\varepsilon}}{(4\pi)^{d/2}} \Gamma(\varepsilon) - 4\varepsilon \frac{1 + 7\varepsilon - 4\varepsilon^3}{(2-\varepsilon)(1-2\varepsilon)(1+2\varepsilon)} \left(\frac{e^2 m^{-2\varepsilon}}{(4\pi)^{d/2}} \Gamma(\varepsilon) \right)^2 + \cdots
$$

$$
= 1 - \frac{4}{3} \frac{e^2 m^{-2\varepsilon}}{(4\pi)^{d/2}} \Gamma(\varepsilon) - \varepsilon(2 + 15\varepsilon + \cdots) \left(\frac{e^2 m^{-2\varepsilon}}{(4\pi)^{d/2}} \Gamma(\varepsilon) \right)^2 + \cdots
$$

$$(5.60)$$

On the other hand, for the $\overline{\text{MS}}$ charge we have

$$
\frac{e^2(\mu)}{e_0^2} = Z_\alpha^{-1} = 1 + z_1 \frac{\alpha(\mu)}{4\pi\varepsilon} + (z_{20} + z_{21}\varepsilon) \left(\frac{\alpha(\mu)}{4\pi\varepsilon} \right)^2 + \cdots \tag{5.61}
$$

(let's pretend for a moment that we don't know Z_α yet). The ratio $e^2(\mu)/e^2$ must be finite at $\varepsilon \to 0$. At one loop, this requirement gives $z_1 = -4/3$. Setting $\mu = m$, we can substitute

$$
\frac{\alpha(m)}{4\pi\varepsilon} = \frac{e^2 m^{-2\varepsilon}}{(4\pi)^{d/2}} \Gamma(\varepsilon) = \frac{\alpha}{4\pi\varepsilon}
$$

in the two-loop term, because the differences are of a higher order (here $\alpha = e^2/(4\pi)$). Thus we obtain $z_{20} = 0$, $z_{21} = -2$. We have reproduced Z_α^{-1} in $\overline{\text{MS}}$ (4.37), (4.25) from our on-shell calculation. Finally, we arrive at the relation between the $\overline{\text{MS}}$ $\alpha(m)$ and the on-shell α:

$$
\alpha(m) = \alpha \left[1 + 15 \left(\frac{\alpha}{4\pi} \right)^2 + \cdots \right]. \tag{5.62}
$$

The three-loop correction has been calculated by [Broadhurst (1992)]. If we need $\alpha(\mu)$ for $\mu \neq m$, we can solve the RG equation (2.77) with this initial condition.

5.8   On-shell renormalization in QCD

QCD perturbation theory is only applicable at large momenta (or small distances). Therefore, it makes no sense to renormalize the gluon field, the coupling, as well as light-quark masses and fields, in the on-shell scheme.

However, it is possible (and often convenient) to use this scheme for renormalizing heavy-quark masses and fields, at the same time using $\overline{\text{MS}}$ for $\alpha_s(\mu)$, the gluon field and light-quark masses and fields. One-loop renormalization constants can be trivially obtained from the QED results (5.29):

$$Z_m^{\text{os}} = Z_Q^{\text{os}} = 1 - C_F \frac{g_0^2 m^{-2\varepsilon}}{(4\pi)^{d/2}} \Gamma(\varepsilon) \frac{d-1}{d-3}. \tag{5.63}$$

Similarly, the relation between the $\overline{\text{MS}}$ mass $m(m)$ and the on-shell mass m is, from (5.31),

$$\frac{m(m)}{m} = 1 - 4C_F \frac{\alpha_s(m)}{4\pi} + \cdots \tag{5.64}$$

The two-loop correction has been found by [Gray *et al.* (1990)] and the three-loop one — by [Melnikov and van Ritbergen (2000)][1]. The quark-field renormalization constant Z_Q^{os} at two loops [Broadhurst *et al.* (1991)] is gauge-invariant, but at three loops [Melnikov and van Ritbergen (2000)] not (unlike in QED).

---

[1]It had been found numerically [Chetyrkin and Steinhauser (1999)] before the analytical result [Melnikov and van Ritbergen (2000)] was obtained.

Chapter 6

Decoupling of heavy-particle loops

6.1 Photon field

Let's consider QED with massless electrons and muons having mass M. When we consider processes with characteristic energies $E \ll M$, the existence of muons is not important. Everything can be described by an effective low-energy theory, in which there are no muons. In other words, muons only exist in loops of size $\sim 1/M$; if we are interested in processes having characteristic distances much larger than $1/M$, such loops can be replaced by local interactions of electrons and photons.

The effective low-energy theory contains the light fields — electrons and photons. The Lagrangian of this theory, describing interactions of these fields at low energies, contains all operators constructed from these fields which are allowed by the symmetries. Operators with dimensionalities $d_i > 4$ are multiplied by coefficients having negative dimensionalities; these coefficients contain $1/M^{d_i-4}$. Therefore, this Lagrangian can be viewed as an expansion in $1/M$. The coefficients in this Lagrangian are fixed by matching — equating S-matrix elements up to some power of p_i/M.

Operators of the full theory are also expansions in $1/M$, in terms of all operators of the effective theory with appropriate quantum numbers. In particular, the bare electron and the photon fields of the full theory are, up to $1/M^2$ corrections,

$$\psi_0 = \left(\zeta_\psi^0\right)^{1/2} \psi_0', \quad A_0 = \left(\zeta_A^0\right)^{1/2} A_0' \qquad (6.1)$$

(primed quantities are those in the effective theory). The bare parameters in the Lagrangians of the two theories are related by

$$e_0 = \left(\zeta_\alpha^0\right)^{1/2} e_0', \quad a_0 = \zeta_A^0 a_0'. \qquad (6.2)$$

The $\overline{\text{MS}}$ renormalized fields and parameters are related by

$$\psi(\mu) = \zeta_\psi^{1/2}(\mu)\psi'(\mu), \quad A(\mu) = \zeta_A^{1/2}(\mu)A'(\mu),$$
$$\alpha(\mu) = \zeta_\alpha(\mu)\alpha'(\mu), \quad a(\mu) = \zeta_A(\mu)a'(\mu), \tag{6.3}$$

where

$$\zeta_\psi(\mu) = \frac{Z'_\psi(\alpha'(\mu), a'(\mu))}{Z_\psi(\alpha(\mu), a(\mu))}\zeta_\psi^0, \quad \zeta_A(\mu) = \frac{Z'_A(\alpha'(\mu))}{Z_A(\alpha(\mu))}\zeta_A^0,$$
$$\zeta_\alpha(\mu) = \frac{Z'_\alpha(\alpha'(\mu))}{Z_\alpha(\alpha(\mu))}\zeta_\alpha^0. \tag{6.4}$$

The photon propagators in the two theories are related by

$$D_\perp(p^2)\left(g_{\mu\nu} - \frac{p_\mu p_\nu}{p^2}\right) + a_0 \frac{p_\mu p_\nu}{(p^2)^2}$$
$$= \zeta_A^0\left[D'_\perp(p^2)\left(g_{\mu\nu} - \frac{p_\mu p_\nu}{p^2}\right) + a'_0 \frac{p_\mu p_\nu}{(p^2)^2}\right] + \mathcal{O}\left(\frac{1}{M^2}\right). \tag{6.5}$$

This explains why the same decoupling constant ζ_A describes decoupling for both the photon field A and the gauge-fixing parameter a. It is most convenient to do matching at $p^2 \to 0$, then the power-suppressed terms in (6.5) play no role. The full-theory propagator near the mass shell is

$$D_\perp(p^2) = \frac{Z_A^{\text{os}}}{p^2}, \quad Z_A^{\text{os}} = \frac{1}{1 - \Pi(0)}. \tag{6.6}$$

Only diagrams with muon loops contribute to $\Pi(0)$, all the other diagrams contain no scale. In the effective theory

$$D'_\perp(p^2) = \frac{Z_A'^{\text{os}}}{p^2}, \quad Z_A'^{\text{os}} = \frac{1}{1 - \Pi'(0)} = 1, \tag{6.7}$$

because all diagrams for $\Pi'(0)$ vanish. Therefore,

$$\zeta_A^0 = \frac{Z_A^{\text{os}}}{Z_A'^{\text{os}}} = \frac{1}{1 - \Pi(0)}. \tag{6.8}$$

At one loop, from (2.32),

$$Z_A(\alpha) = 1 - \frac{8}{3}\frac{\alpha}{4\pi\varepsilon}, \quad Z'_A(\alpha) = 1 - \frac{4}{3}\frac{\alpha}{4\pi\varepsilon}$$

(because there are two lepton flavours in the full theory). With this accuracy, we may put $\alpha'(\mu) = \alpha(\mu)$. Re-expressing $\Pi(0)$ (5.7) via $\alpha(\mu)$ (and replacing m_0 by the on-shell muon mass M, because the difference is beyond our accuracy), we obtain

$$\zeta_A(\mu) = \frac{Z'_A(\alpha(\mu))}{Z_A(\alpha(\mu))} \frac{1}{1 - \Pi(0)} = 1 - \frac{4}{3} \frac{\alpha(\mu)}{4\pi} \left[\left(\frac{\mu}{M} \right)^{2\varepsilon} e^{\gamma\varepsilon} \Gamma(\varepsilon) - \frac{1}{\varepsilon} \right],$$

and finally

$$\zeta_A(\mu) = 1 - \frac{8}{3} \frac{\alpha}{4\pi} \log \frac{\mu}{M}. \tag{6.9}$$

We can always find the μ-dependence from the RG equation, the initial condition is $\zeta_A(M) = 1$ at one loop.

Let's find $\zeta_A(M)$ with two-loop accuracy. We express ζ_A^0 (6.8) via the renormalized quantities: $\alpha(M)$ and the on-shell muon mass M. Technically, this is nearly the same calculation as in Sect. 5.7. But now we express e_0^2 in the one-loop term via $\alpha(M)$, and $Z_\alpha(\alpha)$ contains two lepton flavours:

$$\zeta_A^0 = 1 - \frac{4}{3} \frac{\alpha(M)}{4\pi\varepsilon} - \left(\frac{16}{9} + 2\varepsilon + 15\varepsilon^2 \right) \left(\frac{\alpha}{4\pi\varepsilon} \right)^2 \tag{6.10}$$

(the extra term $16/9$ as compared to (5.60) comes from the doubled β_0). We can neglect the difference between $\alpha(M)$ and $\alpha'(M)$ with our accuracy (this difference is $\mathcal{O}(\alpha^3)$, see Sect. 6.3). From (4.21) and (4.44) we have

$$\frac{Z'_A(\alpha)}{Z_A(\alpha)} = 1 + \frac{4}{3} \frac{\alpha}{4\pi\varepsilon} + \left(\frac{32}{9} + 2\varepsilon \right) \left(\frac{\alpha}{4\pi\varepsilon} \right)^2.$$

All $1/\varepsilon$ poles cancel in the renormalized decoupling constant $\zeta_A(M)$ (6.4), as they should, and we obtain

$$\zeta_A(M) = 1 - 15 \left(\frac{\alpha(M)}{4\pi} \right)^2 + \cdots \tag{6.11}$$

The RG equation

$$\frac{d \log \zeta_A(\mu)}{d \log \mu} + \gamma_A(\alpha(\mu)) - \gamma'_A(\alpha'(\mu)) = 0 \tag{6.12}$$

can be used to find $\zeta_A(\mu)$ for $\mu \neq M$. For example, for $\mu = M(M)$, the $\overline{\text{MS}}$ muon mass (5.31) normalized at $\mu = M$, the difference

$$M(M) - M = -4 \frac{\alpha}{4\pi} M, \tag{6.13}$$

so that $\zeta_A(M(M))$ is still $1 + \mathcal{O}(\alpha^2)$:

$$\zeta_A(M(M)) = 1 - \frac{13}{3}\left(\frac{\alpha(M)}{4\pi}\right)^2 + \cdots \qquad (6.14)$$

This result can be easily obtained directly: if we use $M_0 = Z_m(\alpha(M))M(M)$ to renormalize the mass in the one-loop term of ζ_A^0 instead of $M_0 = Z_m^{\mathrm{os}}M$, we get the formula similar to (6.10), but with $\alpha(M(M))$ and $13/3$ instead of 15.

6.2   Electron field

The electron propagators in the full theory and in the low-energy theory are related by

$$\not{p}S(p) = \zeta_\psi^0\,\not{p}S'(p) + \mathcal{O}\left(\frac{p^2}{M^2}\right). \qquad (6.15)$$

It is most convenient to do matching at $p \to 0$, where power corrections play no role. The full-theory propagator near the mass shell is

$$S(p) = \frac{Z_\psi^{\mathrm{os}}}{\not{p}}, \quad Z_\psi^{\mathrm{os}} = \frac{1}{1 - \Sigma_V(0)}. \qquad (6.16)$$

Only diagrams with muon loops contribute to $\Sigma_V(0)$, all the other diagrams contain no scale; such diagrams first appear at two loops (Fig. 6.1). In the effective theory

$$S'(p) = \frac{Z_\psi'^{\mathrm{os}}}{\not{p}}, \quad Z_\psi'^{\mathrm{os}} = \frac{1}{1 - \Sigma_V'(0)} = 1, \qquad (6.17)$$

because all diagrams for $\Sigma_V'(0)$ vanish. Therefore,

$$\zeta_\psi^0 = \frac{Z_\psi^{\mathrm{os}}}{Z_\psi'^{\mathrm{os}}} = \frac{1}{1 - \Sigma_V(0)}. \qquad (6.18)$$

At two loops (Fig. 6.1),

$$-i\not{p}\Sigma_V(p^2) = \int \frac{d^dk}{(2\pi)^d} ie_0\gamma^\mu i\frac{\not{k}+\not{p}}{(k+p)^2} ie_0\gamma^\nu \left(\frac{-i}{k^2}\right)^2 i(k^2 g_{\mu\nu} - k_\mu k_\nu)\Pi(k^2), \qquad (6.19)$$

where $i(k^2 g_{\mu\nu} - k_\mu k_\nu)\Pi(k^2)$ is the muon-loop contribution to the photon self-energy (Fig. 5.1). It is transverse; therefore, longitudinal parts of the

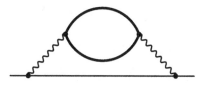

Fig. 6.1 Two-loop diagram for $\Sigma_V(0)$

photon propagators ($\sim \xi k_\alpha k_\beta$) do not contribute, and the result is gauge invariant. We only need the linear term in p in both sides:

$$\not{p}\Sigma_V(0) = -ie_0^2 \int \frac{d^d k}{(2\pi)^d} \gamma^\mu (k^2 \not{p} - 2p\cdot k\,\not{k})\gamma^\nu (k^2 g_{\mu\nu} - k_\mu k_\nu)\frac{\Pi(k^2)}{(k^2)^4}$$

$$= -ie_0^2 \int \frac{d^d k}{(2\pi)^d} \left[\gamma_\mu(k^2\not{p} - 2p\cdot k\,\not{k})\gamma^\mu - \not{k}\not{p}\not{k} + 2p\cdot k\,\not{k} \right] \frac{\Pi(k^2)}{(k^2)^3}.$$

Averaging over k directions by $p\cdot k\,\not{k} \Rightarrow (k^2/d)\not{p}$, we obtain

$$\Sigma_V(0) = -ie_0^2 \frac{(d-1)(d-4)}{d} \int \frac{d^d k}{(2\pi)^d}\frac{\Pi(k^2)}{(k^2)^2}. \tag{6.20}$$

The muon-loop contribution to the photon self-energy (Fig. 5.1) is (we set $M = 1$; the power of M will be restored by dimensionality)

$$i(p^2 g_{\mu\nu} - p_\mu p_\nu)\Pi(p^2) = -\int \frac{d^d k}{(2\pi)^d}\,\mathrm{Tr}\, ie_0\gamma_\mu i\frac{\not{k}+\not{p}+1}{(k+p)^2 - 1}ie_0\gamma_\nu i\frac{\not{k}+1}{k^2 - 1}.$$

Contracting in μ and ν, we obtain

$$\Pi(p^2) = -i\frac{e_0^2}{(d-1)(-p^2)}\int \frac{d^d k}{(2\pi)^d}\frac{N}{D_1 D_2},$$
$$D_1 = 1 - (k+p)^2, \quad D_2 = 1 - k^2, \quad N = \mathrm{Tr}\,\gamma_\mu(\not{k}+\not{p}+1)\gamma^\mu(\not{k}+1).$$

Let's calculate the integral

$$I = \int \frac{d^d p}{(2\pi)^d}\frac{\Pi(p^2)}{(-p^2)^2}, \tag{6.21}$$

which appears in (6.20). It is a two-loop massive vacuum diagram of

Fig. 5.7:

$$I = -i\frac{e_0^2}{d-1}\int\frac{d^dk_1}{(2\pi)^d}\frac{d^dk_2}{(2\pi)^d}\frac{N}{D_1 D_2 D_3^3},$$
$$D_1 = 1 - k_1^2, \quad D_2 = 1 - k_2^2, \quad D_3 = -(k_1 - k_2)^2,$$
$$N = \operatorname{Tr}\gamma_\mu(\not{k}_1 + 1)\gamma^\mu(\not{k}_2 + 1).$$

All scalar products in the numerator can be expressed via the denominators:

$$k_1^2 = 1 - D_1, \quad k_2^2 = 1 - D_2, \quad k_1 \cdot k_2 = \frac{1}{2}(2 + D_3 - D_1 - D_2).$$

Calculating the trace and omitting terms with D_1 or D_2 (which produce vanishing integrals), we obtain

$$N \Rightarrow 2\left[4 - (d-2)D_3\right].$$

Our integral I becomes (see (5.55))

$$I = \frac{2i}{d-1}\frac{e_0^2}{(4\pi)^d}\left[4V(1,1,3) - (d-2)V(1,1,2)\right].$$

Using (5.57) and restoring the power of M, we finally arrive at

$$I = \int\frac{d^dk}{(2\pi)^d}\frac{\Pi(k^2)}{(k^2)^2} = -i\frac{e_0^2 M^{-4\varepsilon}}{(4\pi)^d}\Gamma^2(\varepsilon)\frac{2(d-6)}{(d-2)(d-5)(d-7)}. \tag{6.22}$$

Therefore, we obtain from (6.18)

$$\zeta_\psi^0 = 1 + \frac{e_0^4 M^{-4\varepsilon}}{(4\pi)^d}\Gamma^2(\varepsilon)\frac{2(d-1)(d-4)(d-6)}{d(d-2)(d-5)(d-7)}. \tag{6.23}$$

The renormalized decoupling coefficient is

$$\zeta_\psi(\mu) = \zeta_\psi^0\frac{Z_\psi'(\alpha'(\mu), a'(\mu))}{Z_\psi(\alpha(\mu), a(\mu))}. \tag{6.24}$$

Its μ-dependence can always be found by solving the RG equation. It is sufficient to obtain it at one point, at some specific $\mu \sim M$, to have the initial condition. The most convenient point is $\mu = M$, because $\alpha(M) = \alpha'(M) + \mathcal{O}(\alpha^3)$ (Sect. 6.3) and $a(M) = a'(M) + \mathcal{O}(\alpha^2)$ (Sect. 6.1), and the differences can be neglected with our accuracy. The renormalization constant Z_ψ is given by (4.55), and Z_ψ' — by a similar formula with primed coefficients. Their ratio is

$$\frac{Z_\psi(\alpha, a)}{Z_\psi'(\alpha, a)} = 1 + \frac{1}{4}\left(\gamma_{\psi 0}\Delta\beta_0 + \frac{1}{2}\Delta\gamma_{A0}\gamma_{\psi 0}''a - \Delta\gamma_{\psi 1}\varepsilon\right)\left(\frac{\alpha}{4\pi\varepsilon}\right)^2, \tag{6.25}$$

where $\gamma_{\psi 0}''$ is the coefficient of a in $\gamma_{\psi 0}$, and

$$\Delta\beta_0 = -\frac{4}{3}, \quad \Delta\gamma_{A0} = \frac{8}{3}, \quad \Delta\gamma_{\psi 1} = -4$$

are the single-flavour contributions to β_0, γ_{A0}, $\gamma_{\psi 1}$. We obtain

$$\frac{Z_\psi(\alpha, a)}{Z_\psi'(\alpha, a)} = 1 + \varepsilon \left(\frac{\alpha}{4\pi\varepsilon}\right)^2 .$$

Re-expressing (6.23) via the renormalized $\alpha(M)$,

$$\zeta_\psi^0 = 1 + \varepsilon \left(1 - \frac{5}{6}\varepsilon + \cdots\right) \left(\frac{\alpha}{4\pi\varepsilon}\right)^2 ,$$

we finally obtain

$$\zeta_\psi(M) = \frac{Z_\psi'(\alpha, a)}{Z_\psi(\alpha, a)}\zeta_\psi^0 = 1 - \frac{5}{6}\left(\frac{\alpha(M)}{4\pi}\right)^2 + \cdots \qquad (6.26)$$

The RG equation

$$\frac{d\log\zeta_\psi(\mu)}{d\log\mu} + \gamma_\psi(\alpha(\mu), a(\mu)) - \gamma_\psi'(\alpha'(\mu), a'(\mu)) = 0 \qquad (6.27)$$

can be used to find $\zeta_\psi(\mu)$ for $\mu \neq M$. In contrast to the case of $\zeta_A(\mu)$ (6.12), now $\gamma_\psi - \gamma_\psi'$ is of order α^2, so that changes of μ of order α (such as, e.g., (6.13)) don't change the coefficient of α^2 in (6.26).

6.3 Charge

The proper vertex $e_0\Gamma$ with the external propagators attached (two electron propagators S and one photon propagator D) is the Green function of the fields $\bar\psi_0$, ψ_0, A_0 (i.e., the Fourier transform of the vacuum average of the T-product of these three fields). Therefore, the relation between this quantity in the full theory and in the low-energy effective theory is

$$e_0\Gamma SSD = \zeta_\psi^0 \left(\zeta_A^0\right)^{1/2} e_0'\Gamma'S'S'D' , \qquad (6.28)$$

or, taking into account $S = \zeta_\psi^0 S'$, $D = \zeta_A^0 D'$,

$$e_0\Gamma^\mu = \left(\zeta_\psi^0\right)^{-1} \left(\zeta_A^0\right)^{-1/2} e_0'\Gamma'^\mu \qquad (6.29)$$

In the full theory, the vertex at $p = p'$ on the mass shell ($p^2 = 0$) is

$$\Gamma^\mu = Z_\Gamma^{os}\gamma^\mu . \qquad (6.30)$$

Only diagrams with muon loops contribute to $\Lambda^\mu(p, p)$, all the other diagrams contain no scale; such diagrams first appear at two loops (Fig. 6.2). In the effective theory

$$\Gamma'^\mu = \gamma^\mu \,, \tag{6.31}$$

because all diagrams for $\Lambda'^\mu(p, p)$ vanish. Therefore,

$$\Gamma^\mu = \zeta_\Gamma^0 \Gamma'^\mu \,, \quad \zeta_\Gamma^0 = \frac{Z_\Gamma^{\text{os}}}{Z_\Gamma'^{\text{os}}} = Z_\Gamma^{\text{os}} \,, \tag{6.32}$$

and we obtain from (6.29)

$$\zeta_\alpha^0 = \left(\zeta_\Gamma^0 \zeta_\psi^0 \right)^{-2} \left(\zeta_A^0 \right)^{-1} \,. \tag{6.33}$$

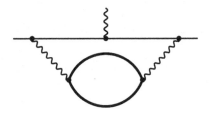

Fig. 6.2   Two-loop diagram for $\Lambda(0, 0)$

The situation in QED is simpler, due to the Ward identity (Sect. 5.4). In the full theory, we have (5.37) $Z_\psi^{\text{os}} Z_\Gamma^{\text{os}} = 1$; similarly, in the effective theory, $Z_\psi'^{\text{os}} Z_\Gamma'^{\text{os}} = 1$ (in fact, these two renormalization constants are equal to 1 separately). Therefore,

$$\zeta_\Gamma^0 \zeta_\psi^0 = 1 \,, \quad \zeta_\alpha^0 = \left(\zeta_A^0 \right)^{-1} \,. \tag{6.34}$$

Recalling also (2.66) $Z_\alpha = Z_A^{-1}$ and $Z_\alpha' = Z_A'^{-1}$, we obtain

$$\zeta_\alpha = \zeta_\alpha^0 \frac{Z_\alpha'}{Z_\alpha} = \left(\zeta_A^0 \frac{Z_A'}{Z_A} \right)^{-1} = \zeta_A^{-1} \,. \tag{6.35}$$

Finally, from (6.11),

$$\zeta_\alpha(M) = 1 + 15 \left(\frac{\alpha(M)}{4\pi} \right)^2 + \cdots \tag{6.36}$$

This means that the running charges in the two theories are related by

$$\alpha(M) = \zeta_\alpha(M)\alpha'(M),$$

i.e., the running charge in the full QED (with both electrons and muons) at $\mu = M$ is slightly larger than in the low-energy effective QED (with only electrons). If you prefer to do the matching at $\mu = M(M)$ instead of $\mu = M$, you should use

$$\zeta_\alpha(M(M)) = 1 + \frac{13}{3}\left(\frac{\alpha(M)}{4\pi}\right)^2 + \cdots \tag{6.37}$$

(see (6.14)).

6.4 Electron mass

In the previous Sections, we considered QED with massless electrons and heavy muons (with mass M). Now let's take the electron mass into account as a small correction. We shall expand everything up to linear terms in m. The electron propagator in the full theory is given by (2.85); in the low-energy effective theory, it involves $\Sigma'_{V,S}$ instead of $\Sigma_{V,S}$. These two propagators are related by ζ_ψ^0:

$$\frac{1}{1 - \Sigma_V(p^2)}\frac{1}{\not{p} - \dfrac{1 + \Sigma_S(p^2)}{1 - \Sigma_V(p^2)}m_0} = \zeta_\psi^0\frac{1}{1 - \Sigma'_V(p^2)}\frac{1}{\not{p} - \dfrac{1 + \Sigma'_S(p^2)}{1 - \Sigma'_V(p^2)}m'_0}. \tag{6.38}$$

Comparing the overall factors, we recover (6.18). Comparing the denominators, we obtain

$$\frac{1 + \Sigma_S(p^2)}{1 - \Sigma_V(p^2)}m_0 = \frac{1 + \Sigma'_S(p^2)}{1 - \Sigma'_V(p^2)}m'_0, \tag{6.39}$$

so that

$$\zeta_m^0 = \frac{m_0}{m'_0} = \frac{1 - \Sigma_V(p^2)}{1 - \Sigma'_V(p^2)}\frac{1 + \Sigma'_S(p^2)}{1 + \Sigma_S(p^2)} = \left(\zeta_\psi^0\right)^{-1}\frac{1}{1 + \Sigma_S(0)}. \tag{6.40}$$

This is because $\Sigma'_S(0) = 0$ in our approximation: after we single out m'_0 in front of $\Sigma'_S(0)$ in the electron self-energy $\Sigma'(p^2)$ (similar to (2.84)), we can set $m'_0 = 0$ in the calculation of $\Sigma'_S(0)$, and it has no scale. Only diagrams with muon loops contribute to $\Sigma_S(0)$; such diagrams first appear at two loops (Fig. 6.1). The renormalized mass decoupling constant is

$$\zeta_m(\mu) = \frac{m(\mu)}{m'(\mu)} = \zeta_m^0 \frac{Z_m'(\alpha'(\mu))}{Z_m(\alpha(\mu))}. \tag{6.41}$$

At two loops (Fig. 6.1), we have to take m_0 into account, as compared to the massless case (6.19):

$$-i\Sigma(p) = \int \frac{d^d k}{(2\pi)^d} ie_0\gamma^\mu i \frac{\not{k} + \not{p} + m_0}{(k+p)^2 - m_0^2} ie_0\gamma^\nu \left(\frac{-i}{k^2}\right)^2 i(k^2 g_{\mu\nu} - k_\mu k_\nu)\Pi(k^2).$$

Again, this is gauge-invariant. In order to extract Σ_S, we have to retain m_0 in the numerator of the electron propagator; after that, we can put $m_0 = 0$:

$$\Sigma_S(0) = -ie_0^2 \int \frac{d^d k}{(2\pi)^d} \frac{\gamma^\mu \gamma^\nu (k^2 g_{\mu\nu} - k_\mu k_\nu)}{(k^2)^3} \Pi(k^2)$$

$$= -ie_0^2(d-1) \int \frac{d^d k}{(2\pi)^d} \frac{\Pi(k^2)}{(k^2)^2}.$$

Using the integral (6.22), we obtain

$$\Sigma_S(0) = -\frac{e_0^4 M^{-4\varepsilon}}{(4\pi)^d} \Gamma^2(\varepsilon) \frac{2(d-1)(d-6)}{(d-2)(d-5)(d-7)}. \tag{6.42}$$

Therefore, from (6.40) and (6.23),

$$\zeta_m^0 = 1 + \frac{e_0^4 M^{-4\varepsilon}}{(4\pi)^d} \Gamma^2(\varepsilon) \frac{8(d-1)(d-6)}{d(d-2)(d-5)(d-7)}. \tag{6.43}$$

It is most easy to calculate $\zeta_m(M)$; $\zeta_m(\mu)$ can then be found by solving the RG equation. To our accuracy, we can take $\alpha(M) = \alpha'(M)$; the ratio of the renormalization constants $Z_m(\alpha)/Z_m'(\alpha)$ is given by the formula similar to (6.25), but γ_m is gauge-invariant ($\gamma_{m0}'' = 0$), and (see (4.58)) $\gamma_{m0} = 6$, $\Delta\gamma_{m1} = -20/3$:

$$\frac{Z_m(\alpha)}{Z_m'(\alpha)} = 1 - \left(2 - \frac{5}{3}\varepsilon\right)\left(\frac{\alpha}{4\pi\varepsilon}\right)^2.$$

Re-expressing (6.43) via the renormalized $\alpha(M)$,

$$\zeta_m^0 = 1 - \left(2 - \frac{5}{3}\varepsilon + \frac{89}{18}\varepsilon^2 + \cdots\right)\left(\frac{\alpha}{4\pi\varepsilon}\right)^2,$$

we finally obtain

$$\zeta_m(M) = 1 - \frac{89}{18}\left(\frac{\alpha(M)}{4\pi}\right)^2 + \cdots \tag{6.44}$$

The RG equation

$$\frac{d \log \zeta_m(\mu)}{d \log \mu} + \gamma_m(\alpha(\mu)) - \gamma'_m(\alpha'(\mu)) = 0 \qquad (6.45)$$

can be used to find $\zeta_m(\mu)$ for $\mu \neq M$ (the difference $\gamma_m - \gamma'_m$ is of order α^2).

6.5 Decoupling in QCD

In QED, effects of decoupling of muon loops are tiny. Also, pion pairs become important at about the same energies as muon pairs, so that QED with electrons and muons is a model with a narrow region of applicability. Therefore, everything we discussed in the previous Sections is not particularly important, from the practical point of view.

In QCD, decoupling of heavy flavours is fundamental and omnipresent. It would be a huge mistake to use the full 6-flavour QCD at characteristic energies of a few GeV, or a few tens of GeV: running of $\alpha_s(\mu)$ and other quantities would be grossly inadequate, convergence of perturbative series would be awful because of huge logarithms. In most cases, anybody working in QCD uses an effective low-energy QCD, where a few heaviest flavours have been removed. Therefore, it is important to understand decoupling in QCD. And to this end the lessons of QED are very helpful.

Suppose we have a heavy flavour with on-shell mass M and n_l light flavours. Then running of the full-theory coupling $\alpha_s^{(n_l+1)}(\mu)$ is governed by the $(n_l + 1)$-flavour β-function; running of the effective-theory coupling $\alpha_s^{(n_l)}(\mu)$ is governed by the n_l-flavour β-function; their matching is given by

$$\alpha_s^{(n_l+1)}(\mu) = \zeta_\alpha(\mu)\alpha_s^{(n_l)}(\mu), \qquad (6.46)$$

with

$$\zeta_\alpha(M) = 1 + \left(15C_F - \frac{32}{9}C_A\right)T_F \left(\frac{\alpha_s(M)}{4\pi}\right)^2 + \cdots \qquad (6.47)$$

Here the C_F term can be obtained from the QED result (6.36) by inserting the obvious colour factors; the C_A term is more difficult to obtain. The RG equation

$$\frac{d \log \zeta_\alpha(\mu)}{d \log \mu} + 2\beta^{(n_l+1)}(\alpha_s^{(n_l+1)}(\mu)) - 2\beta^{(n_l)}(\alpha_s^{(n_l)}(\mu)) = 0 \qquad (6.48)$$

Lectures on QED and QCD

can be used to find $\zeta_\alpha(\mu)$ for $\mu \neq M$. The difference

$$\beta^{(n_l+1)} - \beta^{(n_l)} = -\frac{4}{3}T_F\frac{\alpha_s}{4\pi} + \mathcal{O}(\alpha_s^2),$$

and therefore for the $\overline{\text{MS}}$ mass $M(M)$ (5.64) we have the well-known formula [Larin *et al.* (1995)]

$$\zeta_\alpha(M(M)) = 1 + \left(\frac{13}{3}C_F - \frac{32}{9}C_A\right)T_F\left(\frac{\alpha_s(M)}{4\pi}\right)^2 + \cdots \qquad (6.49)$$

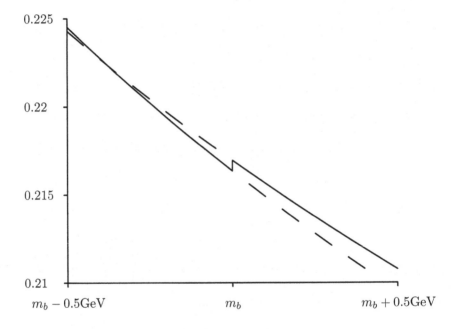

Fig. 6.3 $\quad \alpha_s^{(5)}(\mu)$ and $\alpha_s^{(4)}(\mu)$

The QCD running coupling $\alpha_s(\mu)$ not only runs when μ varies; it also jumps when crossing heavy-flavour thresholds. The behaviour of $\alpha_s(\mu)$ near m_b is shown in Fig. 6.3 (this figure has been obtained using the Mathematica package RunDec [Chetyrkin *et al.* (2000)], which takes into account four-loop β-functions and three-loop decoupling). At $\mu > m_b$, the correct theory is the full 5-flavour QCD ($\alpha_s^{(5)}(\mu)$, the solid line); at $\mu < m_b$, the correct theory is the effective low-energy 4-flavour QCD ($\alpha_s^{(4)}(\mu)$, the solid line); the jump at $\mu = m_b$ (6.47) is shown. Of course, both curves can be continued across m_b (dashed lines), and it is inessential at which particular $\mu \sim m_b$ we switch from one theory to the other one. However, the on-shell mass m_b (or any other mass which differs from it by $\mathcal{O}(\alpha_s)$, such as, e.g., the $\overline{\text{MS}}$ mass $m_b(m_b)$) is most convenient, because the jump is small, $\mathcal{O}(\alpha_s^3)$. For, say, $\mu = 2m_b$ or $\mu = m_b/2$ it would be $\mathcal{O}(\alpha_s^2)$.

Light-quark masses $m_i(\mu)$ also rum with μ, and also jump when crossing a heavy-quark threshold. The QCD result

$$m^{(n_l+1)}(M) = m^{(n_l)}(M) \left[1 - \frac{89}{18} C_F T_F \left(\frac{\alpha_s(M)}{4\pi} \right)^2 + \cdots \right] \qquad (6.50)$$

can be obtained from the QED one (6.44) by inserting the obvious colour factors.

The QCD decoupling relations are currently known at three [Chetyrkin *et al.* (1998)] and even four loops [Chetyrkin *et al.* (2005); Schröder and Steinhauser (2006)].

Chapter 7

Conclusion: Effective field theories

We already discussed effective field theories in Chap. 6. When using QCD, one rarely works with all 6 flavours; more often, one works in an effective low-energy QCD with fewer flavours.

In fact, all our theories (except The Theory of Everything, if such a thing exists) are effective low-energy theories. We don't know physics at arbitrarily small distances (maybe, even our concept of space-time becomes inapplicable at very small distances). We want to describe phenomena at distances larger than some boundary scale; our ignorance about very small distances is parametrized by local interactions of low-mass particles. These observable particles and their interactions are thus described by an effective field theory with all possible local operators in its Lagrangian. Coefficients of higher-dimensional operators have negative dimensionalities, and are proportional to negative powers of the energy scale of a new physics (the scale at which the effective low-energy theory breaks down). At energies much lower than this scale, these higher-dimensional terms in the Lagrangian are unimportant. We may retain dimension-4 terms (renormalizable) and, maybe, one or two power corrections.

The first historical example of an effective low-energy theory is the Heisenberg–Euler effective theory in QED. It is still the best example illustrating typical features of such theories.

In order to understand it better, let's imagine a country, Photonia, in which physicists have high-intensity sources and excellent detectors of low-energy photons, but they don't have electrons and don't know that such a particle exists[1]. At first their experiments (Fig. 7.1a) show that photons do not interact with each other. They construct a theory, Quantum

---

[1]We indignantly refuse to discuss the question "Of what the experimentalists and their apparata are made?" as irrelevant.

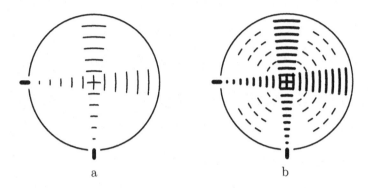

Fig. 7.1 Scattering of low-energy photons

Photodynamics, with the Lagrangian

$$L = -\frac{1}{4} F_{\mu\nu} F^{\mu\nu} . \tag{7.1}$$

But later, after they increased the luminosity (and energy) of their "photon colliders" and the sensitivity of their detectors, they discover that photons do scatter, though with a very small cross-section (Fig. 7.1b). They need to add some interaction terms to this Lagrangian. Lowest-dimensional operators having all the necessary symmetries contain four factors $F_{\mu\nu}$. There are two such terms:

$$L = -\frac{1}{4} F_{\mu\nu} F^{\mu\nu} + c_1 \left(F_{\mu\nu} F^{\mu\nu} \right)^2 + c_2 F_{\mu\nu} F^{\nu\alpha} F_{\alpha\beta} F^{\beta\mu} . \tag{7.2}$$

They can extract the two parameters $c_{1,2}$ from two experimental results, and predict results of infinitely many measurements. So, this effective field theory has predictive power.

We know the underlying more fundamental theory for this effective low-energy theory, namely QED, and so we can help theoreticians from Photonia. The amplitude of photon–photon scattering in QED at low energies must be reproduced by the effective Lagrangian (7.2). At one loop, it is given by the diagram in Fig. 7.2. Expanding it in the photon momenta, we can easily reduce it to the massive vacuum integrals (1.2). Due to the gauge invariance, the leading term is linear in each of the four photon momenta. Then we equate this full-theory amplitude with the effective-theory one following from (7.2), and find the coefficients $c_{1,2}$ (this procedure is

Fig. 7.2   Photon–photon scattering in QED at one loop

known as matching). The result is

$$L = -\frac{1}{4}F_{\mu\nu}F^{\mu\nu} + \frac{\alpha^2}{180m^4}\left[-5\left(F_{\mu\nu}F^{\mu\nu}\right)^2 + 14F_{\mu\nu}F^{\nu\alpha}F_{\alpha\beta}F^{\beta\mu}\right] . \quad (7.3)$$

It is not (very) difficult to calculate two-loop corrections to this QED amplitude using the results of Sect. 5.6, and thus to obtain α^3 terms in these coefficients.

There are many applications of the Lagrangian (7.3). For example, the energy density of the photon gas at temperature T is $\sim T^4$ by dimensionality (Stefan–Boltzmann law). What is the radiative correction to this law? Calculating the vacuum diagram in Fig. 7.3 at temperature T, one can obtain [Kong and Ravndal (1998)] a correction $\sim \alpha^2 T^8/m^4$. Of course, this result is only valid at $T \ll m$.

Fig. 7.3   Radiative correction to the Stefan–Boltzmann law

The interaction terms in the Lagrangian (7.3) contain the "new physics" energy scale, namely the electron mass m, in the denominator. If we want to reproduce more terms in the expansion of QED amplitudes in the ratio ω/m (ω is the characteristic energy), we can include operators of higher dimensions in the effective Lagrangian; their coefficients contain higher powers of m in the denominator. Such operators contain more $F_{\mu\nu}$ and/or its derivatives. Heisenberg and Euler derived the effective Lagrangian for constant field containing all powers of $F_{\mu\nu}$; it is not sufficient for finding coefficients of operators with derivatives of $F_{\mu\nu}$. The expansion in ω/m breaks down when $\omega \sim m$. At such energies the effective low-energy becomes useless, and a more fundamental theory, QED, should be used; in particular, real electron-positron pairs can be produced.

QED, the theory of electrons and photons, is an effective low-energy theory too. Let's forget about hadrons (and τ) for a moment; then the fundamental theory is QED with electrons and muons. Processes with electrons and photons having momenta much lower than the muon mass M can be described by the low-energy effective theory — QED with only electrons:

$$L' = \bar{\psi}'_0 \left(i \slashed{D}' - m'_0 \right) \psi'_0 - \frac{1}{4} F'_{0\mu\nu} F_0'^{\mu\nu} - \frac{1}{2a'_0} \left(\partial_\mu A_0'^\mu \right)^2 \\ + \frac{c_0}{M^2} m'_0 \bar{\psi}'_0 F'_{0\mu\nu} \sigma^{\mu\nu} \psi'_0 + \cdots \tag{7.4}$$

The magnetic-moment interaction (Pauli) operator $\bar{\psi}'_0 F'_{0\mu\nu} \sigma^{\mu\nu} \psi'_0$ has dimensionality 5. If electrons are massless, this operator cannot appear in the effective Lagrangian because of helicity conservation (or chiral symmetry). Therefore, it only appears multiplied by m'_0, the product has dimensionality 6 (there are a few more operators with $d_i = 6$). The coefficient c_0 is obtained by matching, i.e., by calculating the electron magnetic moment in the full theory (with muons), see Fig. 6.2. It is $\sim e^5$; in order to have an even number of γ-matrices, i.e., the helicity flip, we have to retain m_0 in the numerator of one of the electron propagators.

Let's imagine that we only know electrons and photons. There are two ways to search for "new physics":

- To raise energies of our accelerators in the hope to produce real new particles (e.g., muons);
- To measure low-energy quantities (such as the electron magnetic moment) with a high precision in the hope to find effects of higher terms in the effective Lagrangian caused by loops of virtual new particles (e.g., the last term in (7.4)).

This term contains the scale of new physics, the muon mass M, in the denominator. If we find such a deviation from predictions of pure low-energy QED (with only electrons and photons), we can estimate this scale.

In reality, hadronic loops produce a comparable contribution to the electron magnetic moment. This hadronic contribution cannot be calculated theoretically, because no one knows how to calculate low-energy hadronic processes from the first principles of QCD[2]. Therefore, the coefficient of

---

[2]It can be calculated, with a good accuracy, from experimental data about $e^+ e^-$ annihilation into hadrons.

the Pauli term in the QED Lagrangian is a phenomenological parameter, to be extracted from experiment — from the measurement of the electron magnetic moment.

We were lucky that the scale of new physics in QED is far away from the electron mass m. Contributions of heavy-particle loops are also strongly suppressed by powers of α. Therefore, the prediction for the electron magnetic moment from the pure QED Lagrangian (without non-renormalizable corrections) is in good agreement with experiment. After this spectacular success of the simplest Dirac equation (without the Pauli term) for electrons, physicists expected that the same holds for the proton, and its magnetic moment is $e/(2m_p)$. No luck here. This shows that the picture of the proton as a point-like structureless particle is a poor approximation at the energy scale m_p.

Another classical example of low-energy effective theories is the four-fermion theory of weak interactions. It was first proposed by Fermi as a fundamental theory of weak interactions; now it is widely used (in a slightly modified form) as an effective theory of weak interactions at low energies.

Let's consider, for example, b-quark decays. In the Standard Model, they are described by diagrams with W exchange. However, the characteristic energy $E \sim m_b \ll m_W$, and we can neglect the W momentum as compared to m_W, and replace its propagator by the constant $1/m_W^2$. In other words, the W propagation distance $\sim 1/m_W$ is small compared to characteristic distances of the process $\sim 1/E$. We can replace the W exchange by a local four-fermion interaction; it is described by dimension-6 four-fermion operators, their coefficients contain $1/m_W^2$. There are 10 such operators. We can calculate QCD corrections of any order to diagrams containing such four-fermion vertex without any problems; ultraviolet divergences are eliminated by the matrix of renormalization constants of these operators. Their coefficients in the effective low-energy Lagrangian obey a renormalization group equation. The initial condition at $\mu = m_W$ is obtained by matching the full-theory results with the effective-theory ones.

The "new physics" scale where this effective theory breaks down is m_W. As is always the case in effective theories, we have to fix the order of the expansion in E^2/m_W^2 we are interested in. In this case, we work at the first order in $1/m_W^2$, and there can be only one four-fermion vertex in any diagram. What if we want to include the next correction in E^2/m_W^2? In this case we have to include all the relevant dimension-8 operators into the Lagrangian; their coefficients contain $1/m_W^4$. Their number is large but finite. Elimination of ultraviolet divergences of diagrams with two four-fermion vertices (i.e., renormalization of products of dimension-6 operators) requires not only renormalization of these operators separately, but also local overall counterterms (i.e., dimension-8 operators).

We see that the fact that the theory with four-fermion interaction is not renormalizable (as all effective field theories) does not prevent us from calculating finite renormalized results at any fixed order in the expansion in E^2/m_W^2. The number of coefficients in the Lagrangian is finite at each order of this expansion, and the theory retains its predictive power. In practise, this number quickly becomes large, and one can only deal with few power corrections in an effective theory.

In the past, only renormalizable theories were considered well-defined: they contain a finite number of parameters, which can be extracted from a finite number of experimental results and used to predict an infinite number of other potential measurements. Non-renormalizable theories were rejected because their renormalization at all orders in non-renormalizable interactions involve infinitely many parameters, so that such a theory has no predictive power. This principle is absolutely correct, if we are impudent enough to pretend that our theory describes the Nature up to arbitrarily high energies (or arbitrarily small distances).

Our current point of view is more modest. We accept the fact that our theories only describe the Nature at sufficiently low energies (or sufficiently large distances). They are effective low-energy theories. Such theories contain all operators (allowed by the relevant symmetries) in their Lagrangians. They are necessarily non-renormalizable. This does not prevent us from obtaining definite predictions at any fixed order in the expansion in E/M, where E is the characteristic energy and M is the scale of new physics. Only if we are lucky and M is many orders of magnitude larger than the energies we are interested in, we can neglect higher-dimensional operators in the Lagrangian and work with a renormalizable theory.

Practically all physicists believe that the Standard Model is also a low-energy effective theory. But we don't know what is a more fundamental theory whose low-energy approximation is the Standard Model. Maybe, it is some supersymmetric theory (with broken supersymmetry); maybe, it is not a field theory, but a theory of extended objects (superstrings, branes); maybe, this more fundamental theory lives in a higher-dimensional space, with some dimensions compactified; or maybe it is something we cannot imagine at present. The future will tell.

Appendix A

Colour factors

Here we shall discuss how to calculate colour diagrams for $SU(N_c)$ colour group (in the Nature, quarks have $N_c = 3$ colours). For more details, see [Cvitanović (web-book)].

Elements of $SU(N_c)$ are complex $N_c \times N_c$ matrices U which are unitary ($U^+ U = 1$) and have $\det U = 1$. Complex N_c-component column vectors q^i transforming as

$$q \to Uq \quad \text{or} \quad q^i \to U^i{}_j q^j \tag{A.1}$$

form the space in which the fundamental representation operates. Complex conjugate row vectors $q_i^+ = (q^i)^*$ transform as

$$q^+ \to q^+ U^+ \quad \text{or} \quad q_i^+ \to q_j^+ (U^+)^j{}_i \quad \text{where} \quad (U^+)^j{}_i = \left(U^i{}_j\right)^* ; \tag{A.2}$$

this is the conjugated fundamental representation. The scalar product is invariant: $q^+ q' \to q^+ U^+ U q' = q^+ q'$. In other words,

$$\delta^i_j \to \delta^k_l U^i{}_k (U^+)^l{}_j = U^i{}_k (U^+)^k{}_j = \delta^i_j \tag{A.3}$$

is an invariant tensor (it is the colour structure of a meson). The unit antisymmetric tensors $\varepsilon^{i_1 \cdots i_{N_c}}$ and $\varepsilon_{i_1 \dots i_{N_c}}$ are also invariant:

$$\varepsilon^{i_1 \cdots i_{N_c}} \to \varepsilon^{j_1 \cdots j_{N_c}} U^{i_1}{}_{j_1} \cdots U^{i_{N_c}}{}_{j_{N_c}} = \det U \cdot \varepsilon^{i_1 \cdots i_{N_c}} = \varepsilon^{i_1 \cdots i_{N_c}} \tag{A.4}$$

(they are the colour structures of baryons and antibaryons).

Infinitesimal transformations are given by

$$U = 1 + i\alpha^a t^a , \tag{A.5}$$

where α^a are infinitesimal real parameters, and t^a are called generators (of the fundamental representation). They have the following properties:

$$U^+U = 1 + i\alpha^a \left(t^a - (t^a)^+\right) = 1 \;\Rightarrow\; (t^a)^+ = t^a \,,$$
$$\det U = 1 + i\alpha^a \operatorname{Tr} t^a = 1 \qquad\qquad \Rightarrow \; \operatorname{Tr} t^a = 0 \,, \tag{A.6}$$

and are normalized by

$$\operatorname{Tr} t^a t^b = T_F \delta^{ab} \,; \tag{A.7}$$

usually, $T_F = 1/2$ is used, but we shall not specialize it. The space of hermitian matrices is N_c^2-dimensional, and that of traceless hermitian matrices — $(N_c^2 - 1)$-dimensional. Therefore, there are $N_c^2 - 1$ generators t^a which form a basis of this space. Their commutators are i times hermitian traceless matrices, therefore,

$$[t^a, t^b] = i f^{abc} t^c \,, \qquad i f^{abc} = \frac{1}{T_F} \operatorname{Tr}[t^a, t^b] t^c \,, \tag{A.8}$$

where f^{abc} are real constants. Their anticommutators are hermitian; if we subtract their traces, the results will be linear combinations of t^c:

$$[t^a, t^b]_+ = 2\frac{T_F}{N_c} \delta^{ab} + d^{abc} t^c \,, \qquad d^{abc} = \frac{1}{T_F} \operatorname{Tr}[t^a, t^b]_+ t^c \,, \tag{A.9}$$

where d^{abc} are real constants.

The quantities

$$A^a = q^+ t^a q'$$

transform as

$$A^a \to q^+ U^+ t^a U q' = U^{ab} A^b \,; \tag{A.10}$$

this is the adjoint representation. It is defined by

$$U^+ t^a U = U^{ab} t^b \,, \tag{A.11}$$

and hence

$$U^{ab} = \frac{1}{T_F} \operatorname{Tr} U^+ t^a U t^b \,. \tag{A.12}$$

The components $(t^a)^i{}_j$ are some fixed numbers; in other words, they form an invariant tensor (see (A.11)):

$$(t^a)^i{}_j \to U^{ab} U^i{}_k (t^b)^k{}_l (U^+)^l{}_j = (t^a)^i{}_j \,. \tag{A.13}$$

For an infinitesimal transformation,

$$A^a \to q^+(1 - i\alpha^c t^c) t^a (1 + i\alpha^c t^c) q' = q^+(t^a + i\alpha^c i f^{acb} t^b) q',$$

so that

$$U^{ab} = \delta^{ab} + i\alpha^c (t^c)^{ab}, \tag{A.14}$$

where the generators in the adjoint representation are

$$(t^c)^{ab} = i f^{acb}. \tag{A.15}$$

As for any representation, these generators satisfy the commutation relation

$$(t^a)^{dc}(t^b)^{ce} - (t^b)^{dc}(t^a)^{ce} = i f^{abc}(t^c)^{de}; \tag{A.16}$$

it follows from the Jacobi identity

$$[t^a, [t^b, t^d]] + [t^b, [t^d, t^a]] + [t^d, [t^a, t^b]] = 0$$
$$= \left(i f^{bdc} i f^{ace} + i f^{dac} i f^{bce} + i f^{abc} i f^{dce} \right) t^e.$$

It is very convenient to do colour calculations in graphical form [Cvitanović (web-book)]. Quark lines mean δ^i_j, gluon lines mean δ^{ab}, and quark-gluon vertices mean $(t^a)^i_j$. There is no need to invent names for indices; it is much easier to see which indices are contracted — they are connected by a line. Here are the properties of the generators t^a which we already know:

$$\mathrm{Tr}\, 1 = N_c \qquad \text{or} \qquad \bigcirc = N_c,$$

$$\mathrm{Tr}\, t^a = 0 \qquad \text{or} \qquad \bigcirc = 0, \tag{A.17}$$

$$\mathrm{Tr}\, t^a t^b = T_F \delta^{ab} \qquad \text{or} \qquad \bigcirc = T_F \;.$$

There is a simple and systematic method for calculation of colour factors — Cvitanović algorithm [Cvitanović (web-book)]. Now we are going to derive its main identity. The tensor $(t^a)^i_j (t^a)^k_l$ is invariant, because $(t^a)^i_j$ is invariant. It can be expressed via δ^i_j, the only independent invariant tensor with fundamental-representation indices (it is clear that $\varepsilon^{i_1 \cdots i_{N_c}}$ and $\varepsilon_{i_1 \dots i_{N_c}}$ cannot appear in this expression, except the case $N_c = 2$; in this case $\varepsilon^{ik}\varepsilon_{jl}$ can appear, but it is expressible via δ_j^i). The general form of this expression is

$$(t^a)^i{}_j (t^a)^k{}_l = a \left[\delta_l^i \delta_j^k - b \delta_j^i \delta_l^k \right] , \qquad (A.18)$$

or graphically

$$\qquad (A.19)$$

where a and b are some unknown coefficients. If we multiply (A.18) by δ_i^j,

$$(t^a)^i{}_i (t^a)^k{}_l = 0 = a \left[\delta_l^k - b N_c \delta_l^k \right] ,$$

i. e., close the upper line in (A.19),

we obtain

$$b = \frac{1}{N_c} .$$

If we multiply (A.18) by $(t^b)^j{}_i$,

$$(t^b)^j{}_i (t^a)^i{}_j (t^a)^k{}_l = T_F (t^a)^k{}_l = a \left[(t^b)^k{}_l - \frac{1}{N_c}(t^b)^i{}_i \delta_l^k \right] ,$$

i. e., close the upper line in (A.19) onto a gluon,

we obtain

$$a = T_F .$$

The final result is

$$(t^a)^i{}_j(t^a)^k{}_l = T_F \left[\delta^i_l \delta^k_j - \frac{1}{N_c} \delta^i_j \delta^k_l \right] , \qquad (A.20)$$

or graphically

$$\qquad (A.21)$$

This identity allows one to eliminate a gluon exchange in a colour diagram: such an exchange is replaced by the exchange of a quark–antiquark pair, from which its colour-singlet part is subtracted.

It can be also rewritten as

$$\qquad (A.22)$$

or

$$q'^i q_j^+ = \frac{1}{T_F} \left[(q^+ t^a q')(t^a)^i{}_j + \frac{1}{N_c} (q^+ q')\delta^i_j \right] . \qquad (A.23)$$

This shows that the product of the fundamental representation and its conjugate is reducible: it reduces to the sum of two irreducible ones, the adjoint representation and the trivial one. In other words, the state of a quark–antiquark pair with some fixed colours is a superposition of the colour-singlet and the colour-adjoint states.

The Cvitanović algorithm consists of elimination gluon exchanges (A.21) and using simple rules (A.17). Let's consider a simple application: counting gluon colours. Their number is

$$\qquad (A.24)$$

$$= N_c^2 - 1,$$

as we already know.

Now we consider a very important example:

$$= T_F \left(N_c - \frac{1}{N_c} \right) \longrightarrow .$$

The result is

$= C_F \longrightarrow$ or $t^a t^a = C_F ,$ (A.25)

where the Casimir operator in the fundamental representation is

$$C_F = T_F \left(N_c - \frac{1}{N_c} \right) . \tag{A.26}$$

Colour diagrams can contain one more kind of elements: three-gluon vertices if^{abc}. The definition (A.8) when written graphically is

. (A.27)

Let's close the quark line onto a gluon:

Therefore,

(A.28)

This is the final rule of the Cvitanović algorithm: elimination of three-gluon vertices.

The commutation relation (A.16) can be rewritten graphically, similarly to (A.27):

$$(A.29)$$

Sometimes it is easier to use this relation than to follow the Cvitanović algorithm faithfully.

Now we consider another very important example:

$$= 2 \left[\text{} - \frac{1}{N_c} \text{} \right.$$

$$\left. - \text{} + \frac{1}{N_c} \text{} \right]$$

$$= 2 T_F N_c \text{} .$$

The result is

$$\text{} = C_A \text{} \quad \text{or} \quad i f^{acd} i f^{bdc} = C_A \delta^{ab}, \qquad (A.30)$$

where the Casimir operator in the adjoint representation is

$$C_A = 2 T_F N_c. \qquad (A.31)$$

If we want to do calculations with d^{abc} (A.9), we can rewrite this definition graphically:

$$\text{} = \frac{1}{T_F} \left[\text{} + \text{} \right]. \qquad (A.32)$$

We see that d^{abc} (denoted by a small empty circle) is symmetric in all indices. For example, let's find $d^{acd}d^{bdc}$:

$$= \frac{2}{T_F^2} \left[\text{} + \text{} \right]$$

$$= \frac{2}{T_F} \left[\text{} - \frac{1}{N_c} \text{} \right.$$

$$\left. + \text{} - \frac{1}{N_c} \text{} \right]$$

$$= \frac{2}{T_F} \left[\text{} + \text{} \right] - 4\frac{T_F}{N_c} \text{}$$

$$= 2 \left[\text{} - \frac{1}{N_c} \text{} \right.$$

$$\left. + \text{} - \frac{1}{N_c} \text{} \right] - 4\frac{T_F}{N_c} \text{}$$

$$= 2T_F \left(N_c - \frac{4}{N_c} \right) \text{} \, .$$

Another example:

$$= -\frac{T_F}{N_c} \quad \text{} \quad , \tag{A.33}$$

or

$$t^a t^b t^a = -\frac{T_F}{N_c} t^b, \quad -\frac{T_F}{N_c} = C_F - \frac{C_A}{2}. \tag{A.34}$$

One more example:

$$= T_F \left[\quad - \frac{1}{N_c} \right.$$

$$\left. - \quad + \frac{1}{N_c} \right]$$

(A.35)

$$= T_F N_c \qquad ,$$

or

$$i f^{abc} t^b t^a = \frac{C_A}{2} t^c . \tag{A.36}$$

This can be derived in a shorter way. Using antisymmetry of f^{abc}, (A.27) and (A.30),

$$= \frac{1}{2} \left[\quad - \quad \right]$$

(A.37)

$$= \frac{1}{2} \qquad = \frac{C_A}{2}$$

We shall also need the result

$$= \frac{C_A}{2}$$

(A.38)

or $\quad i f^{adf} i f^{bed} i f^{cfe} = \frac{C_A}{2} i f^{abc} .$

It can be, of course, obtained by using the Cvitanović algorithm. A shorter derivation is similar to (A.37), with the lower quark line replaced by the gluon one; the commutation relation (A.29) in the adjoint representation is used instead of (A.27) in the fundamental one.

Sometimes the colour factor of a diagram vanishes, and we don't need to undertake its (maybe, difficult) calculation. For example,

$$\text{(figure)} = 0. \qquad (A.39)$$

This can be obtained by using the Cvitanović algorithm. A shorter derivation is following. There is only one colour structure of the quark–gluon vertex:

$$\text{(figure)} = c \; \text{(figure)} .$$

Therefore, let's close this diagram:

$$c \; \left(\text{figure}\right) = \left(\text{figure}\right) = \left(\text{figure}\right) - \left(\text{figure}\right) = 0.$$

Colour factors of not too complicated diagrams (e.g., propagator and vertex diagrams which have to be calculated for obtaining QCD fields anomalous dimensions and β-function) with L loops (including L_q quark loops) have the form

$$(T_F n_f)^{L_q} C_F^{n_F} C_A^{n_A} , \quad n_F + n_A = L_g = L - L_q . \qquad (A.40)$$

This form is valid not only for $SU(N_c)$, but for any gauge group. At $L_g \geq 4$, new Casimir operators appear, which are not reducible to C_F and C_A. Suppose we have calculated the colour factor for $SU(N_c)$ using the Cvitanović algorithm, and we want to rewrite it in the form (A.40) valid for any group. Our colour factor is $(T_F n_f)^{L_q}$ times a sum of terms of the form $T_F^{L_g} N_c^n$ with various values of n. Using (A.26) and (A.31), we can rewrite each such term as

$$\left(\frac{C_A}{2}\right)^x \left(\frac{C_A}{2} - C_F\right)^y \quad \text{where} \quad x + y = L_g , \quad x - y = n .$$

Therefore, we may substitute

$$T_F^{L_g} N_c^n \rightarrow \left[\frac{C_A}{2} \sqrt{1 - \frac{2C_F}{C_A}} \right]^{L_g} \left[\sqrt{1 - \frac{2C_F}{C_A}} \right]^{-n}. \qquad (A.41)$$

Of course, if there are no errors, the result will contain no fractional powers, just a polynomial in C_F and C_A.[3]

---

[3] I am grateful to K.G. Chetyrkin for explaining this trick to me.

Bibliography

D.J. Broadhurst, Z. Phys. C **54** (1992) 599

D.J. Broadhurst, N. Gray, K. Schilcher, Z. Phys. C **52** (1991) 111

K.G. Chetyrkin, Phys. Lett. B **404** (1997) 161

K.G. Chetyrkin, Nucl. Phys. B **710** (2005) 499

K.G. Chetyrkin, B.A. Kniehl, M. Steinhauser, Nucl. Phys. B **510** (1998) 61

K.G. Chetyrkin, J.H. Kühn, M. Steinhauser, Comput. Phys. Commun. **133** (2000) 43

K.G. Chetyrkin, J.H. Kühn, C. Sturm, Nucl. Phys. B **744** (2006) 121

K.G. Chetyrkin, A. Retey, Nucl. Phys. B **583** (2000) 3

K.G. Chetyrkin, M. Steinhauser, Phys. Rev. Lett. **83** (1999) 4001; Nucl. Phys. B **573** (2000) 617

K.G. Chetyrkin, F.V. Tkachov, Nucl. Phys. B **192** (1981) 159

P. Cvitanović, *Group Theory*, http://www.nbi.dk/GroupTheory/

M. Czakon, Nucl. Phys. B **710** (2005) 485

A.I. Davydychev, P. Osland, O.V. Tarasov, Phys. Rev. D **54** (1996) 4087, Erratum: D **59** (1999) 109901

N. Gray, D.J. Broadhurst, W. Grafe, K. Schilcher, Z. Phys. C **48** (1990) 673

D.J. Gross, F. Wilczek, Phys. Rev. Lett. **30** (1973) 1343

G. 't Hooft, unpublished (1971)

I.B. Khriplovich, Sov. J. Nucl. Phys. **10** (1970) 235

X.-w. Kong, F. Ravndal, Nucl. Phys. B **526** (1998) 627

S. Laporta, E. Remiddi, Phys. Lett. B **379** (1996) 283

S.A. Larin, T. van Ritbergen, J.A.M. Vermaseren, Nucl. Phys. B **438** (1995) 278

S.A. Larin, J.A.M. Vermaseren, Phys. Lett. B **303** (1993) 334

K. Melnikov, T. van Ritbergen, Phys. Lett. B **482** (2000) 99; Nucl. Phys. B **591** (2000) 515

M.E. Peskin, D.V. Schroeder, *An Introduction to Quantum Field Theory*, Perseus Books (1995)

H.D. Politzer, Phys. Rev. Lett. **30** (1973) 1346

T. van Ritbergen, J.A.M. Vermaseren, S.A. Larin, Phys. Lett. B **400** (1997) 379

V.A. Smirnov, *Feynman Integral Calculus*, Springer (2006)

Y. Schröder, M. Steinhauser, J. High Energy Phys. **01** (2006) 051

O.V. Tarasov, Preprint JINR P2-82-900, Dubna (1982)

O.V. Tarasov, A.A. Vladimirov, A.Yu. Zharkov, Phys. Lett. B **93** (1980) 429

V.S. Vanyashin, M.V. Terentev, Sov. Phys. JETP **21** (1965) 375

J.A.M. Vermaseren, S.A. Larin, T. van Ritbergen, Phys. Lett. B **405** (1997) 327

A.A. Vladimirov, Theor. Math. Phys. **43** (1980) 417

PART 2

Multiloop calculations

Chapter 8

Massless propagator diagrams

8.1 Introduction

We shall discuss methods of calculation of propagator diagrams in various theories and kinematics, up to three loops. In order to obtain renormalization-group functions, it is often enough to calculate propagator diagrams. Massless diagrams are discussed in Chap. 8; they are useful, e.g., in QCD with light quarks. Heavy Quark Effective Theory (HQET) is discussed in Chap. 9, and massive on-shell diagrams — in Chap. 10. Finally, some mathematical methods, most useful in massless and HQET calculations, are considered in Chap. 11.

Throughout these lectures, we shall discuss scalar Feynman integrals. Tensor integrals can be expanded in tensor structures, scalar coefficients are found by solving linear systems. Their solution can be written as projectors (tensor, γ-matrix) applied to the original diagram.

Scalar products of momenta in the numerator can be expressed via the denominators. In some classes of problems, there are not enough independent denominators to express all scalar products. Then we add the required number of scalar products to the basis, and consider scalar integrals with powers of these scalar products in the numerators.

Some diagrams contain insertions into internal lines. The denominators of the propagators on both sides coincide. If powers of these denominators are n_1 and n_2, we can combine them into a single line with the power n_1+n_2 (Fig. 8.1).

Throughout these lectures, Minkowski notations with the metric $+---$ are used.

A much more detailed description of many methods for calculating multiloop diagrams can be found in the excellent book [Smirnov (2006)].

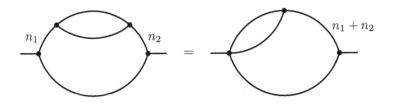

Fig. 8.1   Insertion into a propagator

8.2   One loop

The one-loop massless propagator diagram (Fig. 8.2) is

$$\int \frac{d^d k}{D_1^{n_1} D_2^{n_2}} = i\pi^{d/2}(-p^2)^{d/2-n_1-n_2} G(n_1, n_2) \,, \tag{8.1}$$

$$D_1 = -(k+p)^2 - i0 \,, \qquad D_2 = -k^2 - i0 \,.$$

It is symmetric with respect to $1 \leftrightarrow 2$.

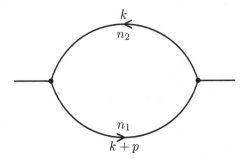

Fig. 8.2   One-loop massless propagator diagram

If n_1 is integer and $n_1 \le 0$, it becomes the vacuum massless diagram (Fig. 8.3) with the polynomial $D_1^{|n_1|}$ in the numerator. It reduces to a sum of powers of $-p^2$ multiplied by integrals

$$\int \frac{d^d k}{(-k^2 - i0)^n} = 0 \,. \tag{8.2}$$

This integral has the mass dimension $d - 2n$; it contains no dimensionful parameters, and the only possible result for it is 0^1. Therefore, $G(n_1, n_2) =$

---

[1]This argument fails at $n = d/2$; more careful investigation [Gorishnii and Isaev (1985)] shows that the right-hand side contains $\delta(d - 2n)$.

0 for non-positive integer n_1 (or n_2). More generally, any massless vacuum diagram with any number of loops is scale-free and hence vanishes.

Fig. 8.3 One-loop massless vacuum diagram

This diagram is calculated by the Fourier transform to x-space and back:

$$\int \frac{e^{-ip\cdot x}}{(-p^2 - i0)^n} \frac{d^d p}{(2\pi)^d} = i2^{-2n}\pi^{-d/2}\frac{\Gamma(d/2 - n)}{\Gamma(n)} \frac{1}{(-x^2 + i0)^{d/2-n}}, \quad (8.3)$$

$$\int \frac{e^{ip\cdot x}}{(-x^2 + i0)^n} d^d x = -i2^{d-2n}\pi^{d/2}\frac{\Gamma(d/2 - n)}{\Gamma(n)} \frac{1}{(-p^2 - i0)^{d/2-n}}. \quad (8.4)$$

Our diagram in x-space is the product of two propagators (8.3) with the powers n_1 and n_2:

$$-2^{-2(n_1+n_2)}\pi^{-d}\frac{\Gamma(d/2 - n_1)\Gamma(d/2 - n_2)}{\Gamma(n_1)\Gamma(n_2)} \frac{1}{(-x^2)^{d-n_1-n_2}}.$$

Transforming it back to p-space (8.4), we arrive at

$$G(n_1, n_2) = \frac{\Gamma(-d/2 + n_1 + n_2)\Gamma(d/2 - n_1)\Gamma(d/2 - n_2)}{\Gamma(n_1)\Gamma(n_2)\Gamma(d - n_1 - n_2)}. \quad (8.5)$$

Diagrams are calculated by going to Euclidean momentum space: $k_0 = ik_{E0}$, $k^2 = -k_E^2$. Using a dimensionless loop momentum $K = k_E/\sqrt{-p^2}$, we can rewrite (8.1) as

$$\int \frac{d^d K}{[(K + n)^2]^{n_1} [K^2]^{n_2}} = \pi^{d/2}G(n_1, n_2),$$

where n is a unit Euclidean vector ($n^2 = 1$). We can perform inversion

$$K = \frac{K'}{K'^2}, \quad K^2 = \frac{1}{K'^2}, \quad d^d K = \frac{d^d K'}{(K'^2)^d}. \quad (8.6)$$

Then

$$(K + n)^2 = \frac{(K' + n)^2}{K'^2},$$

and we obtain (Fig. 8.4)

$$G(n_1, n_2) = G(n_1, d - n_1 - n_2).\tag{8.7}$$

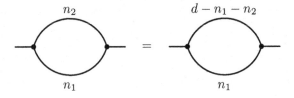

Fig. 8.4   Inversion relation

8.3   Two loops

There is one generic topology of two-loop massless propagator diagrams (Fig. 8.5):

$$\int \frac{d^d k_1\, d^d k_2}{D_1^{n_1} D_2^{n_2} D_3^{n_3} D_4^{n_4} D_5^{n_5}} = -\pi^d (-p^2)^{d - \sum n_i} G(n_1, n_2, n_3, n_4, n_5),$$
$$D_1 = -(k_1 + p)^2, \quad D_2 = -(k_2 + p)^2, \quad D_3 = -k_1^2, \quad D_4 = -k_2^2, \tag{8.8}$$
$$D_5 = -(k_1 - k_2)^2.$$

All other diagrams (e.g., Fig. 8.1) are particular cases of this one, when some line shrinks to a point, i.e., its index $n_i = 0$. This diagram is symmetric with respect to $(1 \leftrightarrow 2, 3 \leftrightarrow 4)$, and with respect to $(1 \leftrightarrow 3, 2 \leftrightarrow 4)$. If indices of any two adjacent lines are non-positive, the diagram contains a scale-free vacuum subdiagram, and hence vanishes.

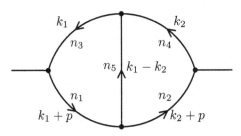

Fig. 8.5   Two-loop massless propagator diagram

If $n_5 = 0$, our diagram is just the product of two one-loop diagrams (Fig. 8.6):

$$G(n_1, n_2, n_3, n_4, 0) = G(n_1, n_3)G(n_2, n_4).$$ (8.9)

If $n_1 = 0$, the inner loop gives $G(n_3, n_5)/(-k_2^2)^{n_3+n_5-d/2}$, and (Fig. 8.7)

$$G(0, n_2, n_3, n_4, n_5) = G(n_3, n_5)G(n_2, n_4 + n_3 + n_5 - d/2).$$ (8.10)

The cases $n_2 = 0$, $n_3 = 0$, $n_4 = 0$ are symmetric.

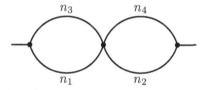

Fig. 8.6   Trivial case $n_5 = 0$

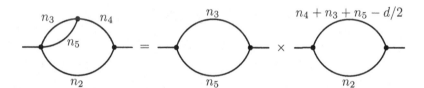

Fig. 8.7   Trivial case $n_1 = 0$

When all $n_i > 0$, the problem does not immediately reduce to a repeated use of the one-loop formula (8.5). We shall use a powerful method called integration by parts [Chetyrkin and Tkachov (1981)]. It is based on the simple observation that any integral of $\partial/\partial k_1(\cdots)$ (or $\partial/\partial k_2(\cdots)$) vanishes (in dimensional regularization no surface terms can appear). From this, we can obtain recurrence relations which involve $G(n_1, n_2, n_3, n_4, n_5)$ with different sets of indices. Applying these relations in a carefully chosen order, we can reduce any $G(n_1, n_2, n_3, n_4, n_5)$ to trivial ones, like (8.9), (8.10).

The differential operator $\partial/\partial k_2$ applied to the integrand of (8.8) acts as

$$\frac{\partial}{\partial k_2} \rightarrow \frac{n_2}{D_2}2(k_2 + p) + \frac{n_4}{D_4}2k_2 + \frac{n_5}{D_5}2(k_2 - k_1).$$ (8.11)

Applying $(\partial/\partial k_2) \cdot k_2$ to the integrand of (8.8), we get a vanishing integral. On the other hand, from (8.8), $2k_2 \cdot k_2 = -2D_4$, $2(k_2 + p) \cdot k_2 = (-p^2) - D_2 - D_4$, $2(k_2 - k_1) \cdot k_2 = D_3 - D_4 - D_5$, we see that this differential operator is equivalent to inserting

$$d - n_2 - n_5 - 2n_4 + \frac{n_2}{D_2}((-p^2) - D_4) + \frac{n_5}{D_5}(D_3 - D_4)$$

under the integral sign (here $(\partial/\partial k_2) \cdot k_2 = d$). Taking into account the definition (8.8), we obtain the recurrence relation

$$\left[d - n_2 - n_5 - 2n_4 + n_2 \mathbf{2^+}(1 - \mathbf{4^-}) + n_5 \mathbf{5^+}(\mathbf{3^-} - \mathbf{4^-})\right] G = 0. \quad (8.12)$$

Here

$$\mathbf{1^{\pm}} G(n_1, n_2, n_3, n_4, n_5) = G(n_1 \pm 1, n_2, n_3, n_4, n_5), \quad (8.13)$$

and similar ones for the other indices.

This is a particular example of the triangle relation. We differentiate in the loop momentum running along the triangle 254, and insert the momentum of the line 4 in the numerator. The differentiation raises the degree of one of the denominators 2, 5, 4. In the case of the line 4, we get $-2D_4$ in the numerator, giving just $-2n_4$. In the case of the line 5, we get the denominator D_3 of the line attached to the vertex 45 of our triangle, minus the denominators D_4 and D_5. The case of the line 2 is similar; the denominator of the line attached to the vertex 24 of our triangle is just $-p^2$, and it does not influence any index of G. Of course, there are three more relations obtained from (8.12) by the symmetry. Another useful triangle relation is derived by applying the operator $(\partial/\partial k_2) \cdot (k_2 - k_1)$:

$$\left[d - n_2 - n_4 - 2n_5 + n_2 \mathbf{2^+}(\mathbf{1^-} - \mathbf{5^-}) + n_4 \mathbf{4^+}(\mathbf{3^-} - \mathbf{5^-})\right] G = 0. \quad (8.14)$$

One more is obtained by the symmetry. Relations of this kind can be written for any diagram having a triangle in it, when at least two vertices of the triangle each have only a single line (not belonging to the triangle) attached.

We can obtain a relation from homogeneity of the integral (8.8) in p. Applying the operator $p \cdot (\partial/\partial p)$ to the integral (8.8), we see that it is equivalent to the factor $2(d - \sum n_i)$. On the other hand, explicit differentiation of the integrand gives $-(n_1/D_1)(-p^2 + D_1 - D_3) - (n_2/D_2)(-p^2 + D_2 - D_4)$.

Therefore,

$$\left[2(d - n_3 - n_4 - n_5) - n_1 - n_2 + n_1 \mathbf{1}^+(1 - \mathbf{3}^-) + n_2 \mathbf{2}^+(1 - \mathbf{4}^-)\right] G = 0.$$
$$(8.15)$$

This is nothing but the sum of the $(\partial/\partial k_2) \cdot k_2$ relation (8.12) and its mirror-symmetric $(\partial/\partial k_1) \cdot k_1$ relation.

Another interesting relation is obtained by inserting $(k_1 + p)^\mu$ into the integrand of (8.8) and taking derivative $\partial/\partial p^\mu$ of the integral. On the one hand, the vector integral must be proportional to p^μ, and we can make the substitution

$$k_1 + p \rightarrow \frac{(k_1 + p) \cdot p}{p^2} p = \left(1 + \frac{D_1 - D_3}{-p^2}\right) \frac{p}{2}$$

in the integrand. Taking $\partial/\partial p^\mu$ of this vector integral produces (8.8) with

$$\left(\tfrac{3}{2}d - \sum n_i\right)\left(1 + \frac{D_1 - D_3}{-p^2}\right)$$

inserted into the integrand. On the other hand, explicit differentiation in p gives

$$d + \frac{n_1}{D_1} 2(k_1 + p)^2 + \frac{n_2}{D_2} 2(k_2 + p) \cdot (k_1 + p),$$
$$2(k_2 + p) \cdot (k_1 + p) = D_5 - D_1 - D_2.$$

Therefore, we obtain

$$\left[\tfrac{1}{2}d + n_1 - n_3 - n_4 - n_5 + \left(\tfrac{3}{2}d - \sum n_i\right)(\mathbf{1}^- - \mathbf{3}^-)\right.$$
$$\left. + n_2 \mathbf{2}^+(\mathbf{1}^- - \mathbf{5}^-)\right] G = 0.$$
$$(8.16)$$

This relation has been derived by S.A. Larin in his M. Sc. thesis. Three more relations follow from the symmetries.

Expressing $G(n_1, n_2, n_3, n_4, n_5)$ from (8.14):

$$G(n_1, n_2, n_3, n_4, n_5) = \frac{n_2 \mathbf{2}^+(\mathbf{5}^- - \mathbf{1}^-) + n_4 \mathbf{4}^+(\mathbf{5}^- - \mathbf{3}^-)}{d - n_2 - n_4 - 2n_5} G, \quad (8.17)$$

we see that the sum $n_1 + n_3 + n_5$ reduces by 1. If we start from an integral belonging to the plane $n_1 + n_3 + n_5 = n$ in Fig. 8.8, then each of the integrals in the right-hand side belong to the plane $n_1 + n_3 + n_5 = n - 1$ (Fig. 8.8). Therefore, applying (8.17) sufficiently many times, we can reduce an arbitrary G integral with integer indices to a combination of integrals

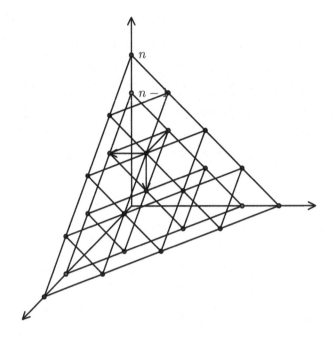

Fig. 8.8   Reduction of $n = n_1 + n_3 + n_5$

with $n_5 = 0$ (Fig. 8.6, (8.9)), $n_1 = 0$ (Fig. 8.7, (8.10)), $n_3 = 0$ (mirror-symmetric to the previous case). Of course, if $\max(n_2, n_4) < \max(n_1, n_3)$, it may be more efficient to use the relation mirror-symmetric to (8.14). The relation (8.16) also can be used instead of (8.14).

Let's summarize. There is one generic topology of two-loop massless propagator diagrams (Fig. 8.9). The integrals with all integer n_i can be expressed as linear combinations of two basis integrals (Fig. 8.10); coefficients are rational functions of d. Here G_n are n-loop massless sunset diagrams

Fig. 8.9   Two-loop massless propagator diagram

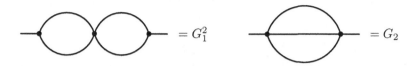

Fig. 8.10   Basis diagrams (all indices equal 1)

(Fig. 8.11):

$$G_n = \frac{1}{\left(n + 1 - n\frac{d}{2}\right)_n \left((n+1)\frac{d}{2} - 2n - 1\right)_n} \frac{\Gamma(1 + n\varepsilon)\Gamma^{n+1}(1 - \varepsilon)}{\Gamma(1 - (n+1)\varepsilon)} . \quad (8.18)$$

Fig. 8.11 n-loop massless sunset diagrams

Using inversion (8.6) of both loop momenta and

$$(K_1 - K_2)^2 = \frac{(K_1' - K_2')^2}{K_1'^2 K_2'^2} ,$$

we obtain relation of Fig. 8.12.[2]

---

[2]This is one of the elements of the symmetry group $Z_2 \times S_6$ [Broadhurst (1986); Barfoot and Broadhurst (1988)] (with 1440 elements) of the diagram of Fig. 8.9. If we connect the external vertices by a line having $n_6 = \frac{3}{2}d - \sum_{i=1}^{5} n_i$, we obtain a logarithmically divergent three-loop tetrahedron vacuum diagram. It is proportional to our original diagram times the logarithm of the ultraviolet cut-off. We can cut it at any of its lines, and get a diagram of Fig. 8.9 with 5 indices n_i out of 6. This gives the tetrahedron symmetry group. Also, we have the inversion relations (Fig. 8.12). Fourier transform to x-space gives a new integral of the form of Fig. 8.9 with $n_i \to \frac{d}{2} - n_i$. There is an additional symmetry following from the star-triangle relation [Kazakov (1984); Kazakov (1985)].

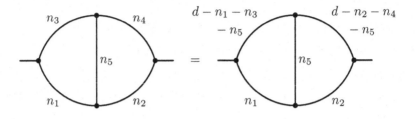

Fig. 8.12   Inversion relation

8.4   Three loops

There are three generic topologies of three-loop massless propagator diagrams (Fig. 8.13). Each has 8 denominators. There are 9 scalar products of three loop momenta k_i and the external momentum p. Therefore, for each topology, all scalar products in the numerator can be expressed via the denominators and one selected scalar product. Integration-by-parts recurrence relations for these diagrams have been investigated by [Chetyrkin and Tkachov (1981)]. They can be used to reduce all integrals of Fig. 8.13, with arbitrary integer powers of denominators and arbitrary (non-negative)

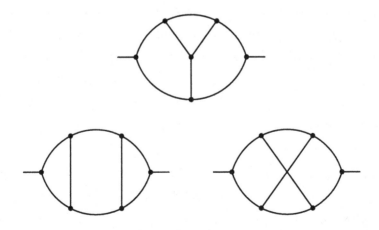

Fig. 8.13   Topologies of three-loop massless propagator diagrams

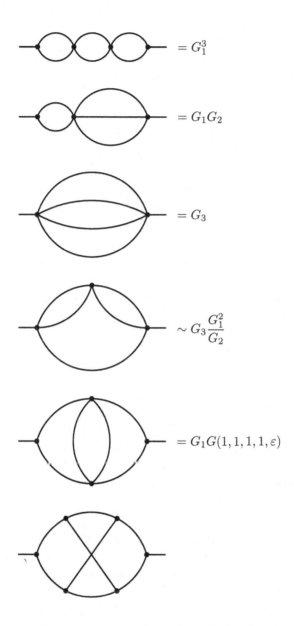

Fig. 8.14   Basis diagrams (all indices equal 1, no numerators)

powers of the selected scalar product in the numerators, to linear combinations of 6 basis integrals (Fig. 8.14). This algorithm has been implemented in the SCHOONSCHIP [Veltman (1967)] package MINCER [Gorishny *et al.* (1984)] and later re-implemented [Larin *et al.* (1991)] in FORM [Vermaseren (1991)]. It has also been implemented in the REDUCE [Hearn (1999); Grozin (1997)] package Slicer [Broadhurst (1992a)]. Only the last, non-planar, topology in Fig. 8.13 involves the last, non-planar, basis integral in Fig. 8.14.

The first four basis integrals are trivial: they are expressed via G_n (8.18) (Fig. 8.11), and hence via Γ-functions. The fourth one differs from the third one (G_3) by replacing the two-loop subdiagram: the second one in Fig. 8.10 ($G_2/(-k^2)^{3-d}$) by the first one ($G_1^2/(-k^2)^{4-d}$). Therefore, it can be obtained from G_3 by multiplying by

$$\frac{G_1^2 G(1, 4 - d)}{G_2 G(1, 3 - d)} = \frac{2d - 5}{d - 3} \frac{G_1^2}{G_2}.$$

The fifth basis integral is proportional to the two-loop diagram $G(1, 1, 1, 1, n)$ with a non-integer index of the middle line $n = \varepsilon$, and will be discussed in Sect. 8.5. The sixth one, non-planar, is truly three-loop and most difficult; it will be discussed in Sect. 8.6.

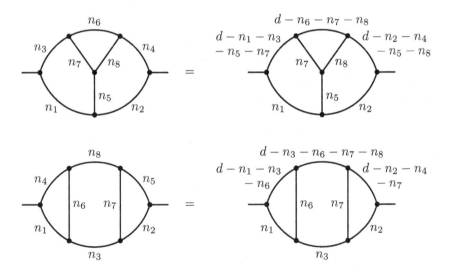

Fig. 8.15 Inversion relations

Performing inversion (8.6) of the loop momenta, we obtain the relations in Fig. 8.15. For example, the ladder diagram with all indices $n_i = 1$ is convergent; its value at $d = 4$ is related to a simpler diagram (Fig. 8.16) by the second inversion relation. The non-planar topology (Fig. 8.13) involves lines with sums of three momenta; they don't transform into anything reasonable under inversion, and there is no inversion relation for it.

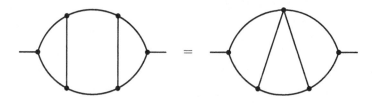

Fig. 8.16   Inversion relation: all $n_i = 1$, $d = 4$

8.5 $G(1, 1, 1, 1, n)$

This diagram is [Kotikov (1996)]

$$G(1,1,1,1,n) = 2\Gamma\left(\tfrac{d}{2} - 1\right)\Gamma\left(\tfrac{d}{2} - n - 1\right)\Gamma(n - d + 3) \times$$
$$\left[\frac{2\Gamma\left(\tfrac{d}{2} - 1\right)}{(d - 2n - 4)\Gamma(n+1)\Gamma\left(\tfrac{3}{2}d - n - 4\right)} {}_3F_2\left(\begin{array}{c} 1, d - 2, n - \tfrac{d}{2} + 2 \\ n + 1, n - \tfrac{d}{2} + 3 \end{array}\middle| 1\right)\right.$$
$$\left. - \frac{\pi \cot \pi(d - n)}{\Gamma(d - 2)}\right] \tag{8.19}$$

This is a particular case of a more general result [Kotikov (1996)] for the diagram with three non-integer indices. The two lines with unit indices must be adjacent (Fig. 8.17); all such diagrams are equivalent, due to the tetrahedron symmetry mentioned above. This result was derived using Gegenbauer polynomial technique [Chetyrkin *et al.* (1980)]; we shall discuss it for a simpler example in Sect. 9.6. When n is integer, the ${}_3F_2$ in (8.19) is expressed via Γ-functions; in order to show equivalence to the standard results (Sect. 8.3), one should use

$$\Gamma(1 + l\varepsilon)\Gamma(1 - l\varepsilon) = \frac{\pi l\varepsilon}{\sin \pi l\varepsilon}$$

and trigonometric identities.

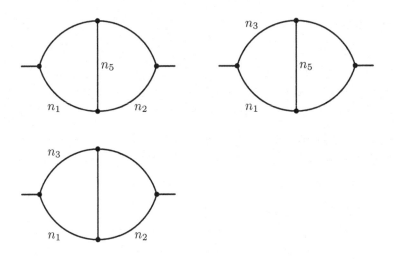

Fig. 8.17   Diagrams calculated in [Kotikov (1996)]

Similar expressions for these diagrams were also obtained in [Broadhurst *et al.* (1997)][3]. In our particular case, they reduce to

$$\frac{(d-3)(d-4)\Gamma(n)\Gamma\left(\frac{3}{2}d-n-4\right)}{2\Gamma(n+3-d)\Gamma^2\left(\frac{d}{2}-1\right)\Gamma\left(\frac{d}{2}-n-1\right)}G(1,1,1,1,n)$$

$$=\frac{3d-2n-10}{d-n-3}\,_3F_2\left(\begin{array}{c}1,\frac{d}{2}-2,n+3-d\\n,n+4-d\end{array}\middle|1\right)$$

$$+\frac{\Gamma(n)\Gamma\left(\frac{3}{2}d-n-4\right)}{\Gamma(d-4)\Gamma\left(\frac{d}{2}-1\right)}\pi\cot\pi(n-d)-2\Gamma\left(\frac{d}{2}-1\right)$$

$$=-\frac{3d-2n-10}{d-n-3}\,_3F_2\left(\begin{array}{c}1,1-n,d-n-3\\3-\frac{d}{2},d-n-2\end{array}\middle|1\right)$$

$$+\frac{\Gamma(n)\Gamma\left(\frac{3}{2}d-n-4\right)}{\Gamma(d-4)\Gamma\left(\frac{d}{2}-1\right)}\pi\cot\pi\frac{d}{2}+\frac{d-4}{d-n-3}\Gamma\left(\frac{d}{2}-1\right)$$

$$=4\frac{n-1}{d-2n-2}\,_3F_2\left(\begin{array}{c}1,n+5-\frac{3}{2}d,n+1-\frac{d}{2}\\3-\frac{d}{2},n+2-\frac{d}{2}\end{array}\middle|1\right)$$

$$+\frac{\Gamma(n)\Gamma\left(\frac{3}{2}d-n-4\right)}{\Gamma(d-4)\Gamma\left(\frac{d}{2}-1\right)}\pi\cot\pi\frac{d}{2}-2\frac{d-4}{d-2n-2}\Gamma\left(\frac{d}{2}-1\right)$$

---

[3]The statement that these diagrams can be expressed via $_3F_2$ functions of unit argument was communicated by D. Broadhurst to A. Kotikov before [Kotikov (1996)].

$$= -4\frac{n-1}{d-2n-2}{}_3F_2\left(\begin{matrix}1,\frac{d}{2}-2,\frac{d}{2}-n-1\\\frac{3}{2}d-n-4,\frac{d}{2}-n\end{matrix}\middle|1\right)$$

$$+ \frac{\Gamma(n)\Gamma\left(\frac{3}{2}d-n-4\right)}{\Gamma(d-4)\Gamma\left(\frac{d}{2}-1\right)}\pi\cot\pi\left(\frac{d}{2}-n\right)-2\Gamma\left(\frac{d}{2}-1\right).$$

(8.20)

[Kazakov (1985)] obtained an expression for $G(1,1,1,1,n)$ via two ${}_3F_2$ functions of argument -1 much earlier. Recently, $G(n_1, n_2, n_3, n_4, n_5)$ for arbitrary indices was calculated [Bierenbaum and Weinzierl (2003)] in terms of a double integral or several double series. These series can be systematically expanded in ε up to any order, and coefficients are expressed via multiple ζ-values (Sect. 11.1).

In order to calculate the two-loop diagram of Fig. 8.9 with a one-loop insertion in the middle light line, we need $G(1,1,1,1,n+\varepsilon)$. It is easy to shift n_5 by ± 1 using the relation [Chetyrkin and Tkachov (1981)]

$$\left[(d-2n_5-4)\mathbf{5}^+ + 2(d-n_5-3)\right]G(1,1,1,1,n_5)$$
$$= 2\,\mathbf{1}^+(\mathbf{3}^- - \mathbf{2}^-\mathbf{5}^+)G(1,1,1,1,n_5),$$

(8.21)

which follows from integration-by-parts relations (Sect. 8.3) (all terms in its right-hand side are trivial). Therefore, it is sufficient to find it just for one n. The simplest choice for which the algorithm of ε-expansion (Chap. 11) works for (8.19) is $n_5 = 2 + \varepsilon$:

$$G(1,1,1,1,2+\varepsilon) = \frac{2\Gamma(1+3\varepsilon)\Gamma(1-\varepsilon)}{1+2\varepsilon}\times$$

$$\left[\frac{\Gamma(1-2\varepsilon)\Gamma(1-\varepsilon)}{(2+\varepsilon)(1+\varepsilon)^2\Gamma(1-4\varepsilon)\Gamma(1+\varepsilon)}{}_3F_2\left(\begin{matrix}1,2-2\varepsilon,2+2\varepsilon\\3+\varepsilon,3+2\varepsilon\end{matrix}\middle|1\right)\right.$$

$$\left.+\frac{\pi\cot 3\pi\varepsilon}{2\varepsilon(1-2\varepsilon)}\right].$$

(8.22)

Expansion of this ${}_3F_2$ in ε is (Chap. 11)

$${}_3F_2\left(\begin{matrix}1,2-2\varepsilon,2+2\varepsilon\\3+\varepsilon,3+2\varepsilon\end{matrix}\middle|1\right) = 4(\zeta_2-1)+6(-4\zeta_3+3\zeta_2-1)\varepsilon$$

$$+ 2(41\zeta_4-54\zeta_3+22\zeta_2-9)\varepsilon^2$$

$$+ 3(-124\zeta_5+24\zeta_2\zeta_3+123\zeta_4-88\zeta_3+30\zeta_2-8)\varepsilon^3+\cdots$$

(8.23)

Similarly, we obtain from (8.20) for $n_5 = 1 + \varepsilon$

$$\frac{3\varepsilon^3(1-2\varepsilon)\Gamma(1+\varepsilon)\Gamma(1-4\varepsilon)}{\Gamma^2(1-\varepsilon)\Gamma(1-2\varepsilon)\Gamma(1+3\varepsilon)}G(1,1,1,1,1+\varepsilon)$$

$$= \frac{4}{3}\,{}_3F_2\left(\begin{array}{c}1,-\varepsilon,3\varepsilon\\1+\varepsilon,1+3\varepsilon\end{array}\bigg|\,1\right) - \frac{\Gamma(1+\varepsilon)\Gamma(1-4\varepsilon)}{\Gamma(1-\varepsilon)\Gamma(1-2\varepsilon)}\pi\varepsilon\cot 3\pi\varepsilon - 1$$

$$= -\frac{4}{3}\,{}_3F_2\left(\begin{array}{c}1,-\varepsilon,-3\varepsilon\\1+\varepsilon,1-3\varepsilon\end{array}\bigg|\,1\right) + \frac{\Gamma(1+\varepsilon)\Gamma(1-4\varepsilon)}{\Gamma(1-\varepsilon)\Gamma(1-2\varepsilon)}\pi\varepsilon\cot \pi\varepsilon + \frac{1}{3} \quad (8.24)$$

$$= -\frac{1}{2}\,{}_3F_2\left(\begin{array}{c}1,4\varepsilon,2\varepsilon\\1+\varepsilon,1+2\varepsilon\end{array}\bigg|\,1\right) + \frac{\Gamma(1+\varepsilon)\Gamma(1-4\varepsilon)}{\Gamma(1-\varepsilon)\Gamma(1-2\varepsilon)}\pi\varepsilon\cot \pi\varepsilon - \frac{1}{2}$$

$$= \frac{1}{2}\,{}_3F_2\left(\begin{array}{c}1,-\varepsilon,-2\varepsilon\\1-4\varepsilon,1-2\varepsilon\end{array}\bigg|\,1\right) + \frac{\Gamma(1+\varepsilon)\Gamma(1-4\varepsilon)}{\Gamma(1-\varepsilon)\Gamma(1-2\varepsilon)}\pi\varepsilon\cot 2\pi\varepsilon - 1\,,$$

where

$$ {}_3F_2\left(\begin{array}{c}1,-\varepsilon,3\varepsilon\\1+\varepsilon,1+3\varepsilon\end{array}\bigg|\,1\right) = 1 - 3\zeta_2\varepsilon^2 + 18\zeta_3\varepsilon^3 - \frac{123}{2}\zeta_4\varepsilon^4$$

$$+ 9(31\zeta_5 - 6\zeta_2\zeta_3)\varepsilon^5 + \cdots$$

$$ {}_3F_2\left(\begin{array}{c}1,-\varepsilon,-3\varepsilon\\1+\varepsilon,1-3\varepsilon\end{array}\bigg|\,1\right) = 1 + 3\zeta_2\varepsilon^2 + \frac{69}{2}\zeta_4\varepsilon^4 + 27(-3\zeta_5 + 2\zeta_2\zeta_3)\varepsilon^5 + \cdots$$

$$ {}_3F_2\left(\begin{array}{c}1,4\varepsilon,2\varepsilon\\1+\varepsilon,1+2\varepsilon\end{array}\bigg|\,1\right) = 1 + 8\zeta_2\varepsilon^2 + 92\zeta_4\varepsilon^4 + 72(-3\zeta_5 + 2\zeta_2\zeta_3)\varepsilon^5 + \cdots$$

$$ {}_3F_2\left(\begin{array}{c}1,-\varepsilon,-2\varepsilon\\1-4\varepsilon,1-2\varepsilon\end{array}\bigg|\,1\right) = 1 + 2\zeta_2\varepsilon^2 + 18\zeta_3\varepsilon^3 + 101\zeta_4\varepsilon^4$$

$$+ 18(23\zeta_5 + 2\zeta_2\zeta_3)\varepsilon^5 + \cdots \quad (8.25)$$

Also, the last formula in (8.20) gives for $n_5 = \varepsilon$

$$G(1,1,1,1,\varepsilon) = \frac{2\Gamma(1-\varepsilon)\Gamma(1+3\varepsilon)}{3\varepsilon(1-2\varepsilon)(1-3\varepsilon)(1-4\varepsilon)\Gamma(1+\varepsilon)\Gamma(1-4\varepsilon)}$$

$$\times\left[\frac{1-\varepsilon}{1-2\varepsilon}\Gamma(1-\varepsilon)\Gamma(1-2\varepsilon){}_3F_2\left(\begin{array}{c}1,-\varepsilon,1-2\varepsilon\\2-4\varepsilon,2-2\varepsilon\end{array}\bigg|\,1\right)\right. \quad (8.26)$$

$$\left.+ (1-4\varepsilon)\Gamma(1+\varepsilon)\Gamma(1-4\varepsilon)\pi\cot 2\pi\varepsilon - \Gamma(1-\varepsilon)\Gamma(1-2\varepsilon)\right],$$

where

$$
{}_3F_2\left(\begin{array}{c} 1, -\varepsilon, 1-2\varepsilon \\ 2-4\varepsilon, 2-2\varepsilon \end{array}\middle| 1\right) = 1 + (\zeta_2 - 2)\varepsilon + (9\zeta_3 - 5\zeta_2 - 3)\varepsilon^2
$$
$$
+ \left(\frac{101}{2}\zeta_4 - 45\zeta_3 + 3\zeta_2 - 6\right)\varepsilon^3 \tag{8.27}
$$
$$
+ \left(207\zeta_5 + 18\zeta_2\zeta_3 - \frac{505}{2}\zeta_4 + 27\zeta_3 + 3\zeta_2 - 15\right)\varepsilon^4 + \cdots
$$

It is not difficult to find several additional terms to (8.23), (8.25), (8.27). Of course, all 4 series (8.24) for $G(1,1,1,1,1+\varepsilon)$ agree, and those for $G(1,1,1,1,2+\varepsilon)$ (8.22) and $G(1,1,1,1,\varepsilon)$ (8.26) agree with the recurrence relation (8.21).

8.6 Non-planar basis integral

The non-planar basis diagram (the last one in Fig. 8.14) is finite; its value at $\varepsilon = 0$ can be obtained without calculations by the method called gluing [Chetyrkin and Tkachov (1981)].

Let's consider the four-loop vacuum diagram in the middle of Fig. 8.18. Let all lines have mass m, then there are no infrared divergences. There are no divergent subdiagrams; the diagram has an overall ultraviolet divergence, and hence a $1/\varepsilon$ pole.

We can imagine that the middle vertex consists of two vertices connected by a shrunk line (with power 0), one vertex involves two lines on the left, and the other one — two lines on the right. If we cut this shrunk line, we get the ladder diagram (the upper one in Fig. 8.18). It is finite. Therefore, the ultraviolet divergence of the vacuum diagram comes from the last integration in $d^d p$ of the ladder diagram. At $p \to \infty$, we may neglect m; by dimensionality, the massless ladder diagram behaves as $(A + \mathcal{O}(\varepsilon))/(-p^2)^{2+3\varepsilon}$. Reducing it to the basis integrals and using results of Sect. 8.5, we can obtain $A = 20\zeta_5$ (the first line of Fig. 8.18)[4]. It is not difficult to obtain several additional terms. The ultraviolet divergence of

---

[4]this $\mathcal{O}(1)$ term was found by [Chetyrkin and Tkachov (1981)] using Gegenbauer polynomials in x-space [Chetyrkin *et al.* (1980)]; however, it is difficult to obtain further terms of ε expansion by this method.

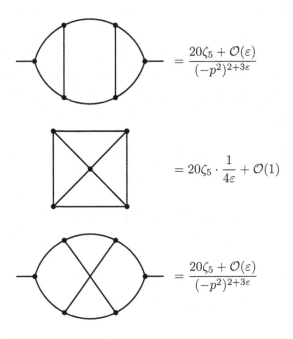

$$= \frac{20\zeta_5 + \mathcal{O}(\varepsilon)}{(-p^2)^{2+3\varepsilon}}$$

$$= 20\zeta_5 \cdot \frac{1}{4\varepsilon} + \mathcal{O}(1)$$

$$= \frac{20\zeta_5 + \mathcal{O}(\varepsilon)}{(-p^2)^{2+3\varepsilon}}$$

Fig. 8.18   Gluing

the vacuum diagram comes from

$$-\frac{i}{\pi^{d/2}} \int \frac{d^d p}{(-p^2)^{2+3\varepsilon}}\bigg|_{\mathrm{UV}} = \frac{2}{\Gamma(2-\varepsilon)} \int_\lambda^\infty p_E^{-1-8\varepsilon} dp_E$$

$$= \frac{2}{\Gamma(2-\varepsilon)} \frac{\lambda^{-8\varepsilon}}{8\varepsilon} = \frac{1}{4\varepsilon} + \mathcal{O}(1),$$

where λ is an infrared cutoff, see the middle line in Fig. 8.18.

On the other hand, we can imagine that one of the vertices in the middle of the vacuum diagram involves the upper left line and the lower right one, and the other vertex — the lower left line and the upper right one. Cutting the shrunk line, we get the non-planar diagram (the lower one in Fig. 8.18). It is finite; at $p \to \infty$, where m can be neglected, it behaves as $(B+\mathcal{O}(\varepsilon))/(-p^2)^{2+3\varepsilon}$. Integrating it in $d^d p$, we obtain the same vacuum diagram, with the same $1/\varepsilon$ pole. Therefore, $B = A$: *the non-planar basis diagram has the same value at $d = 4$ as the ladder one* [Chetyrkin and Tkachov (1981)], namely, $20\zeta_5$ (Fig. 8.18). This method tells us nothing about further terms of expansion in ε. The highly non-trivial problem of calculating the $\mathcal{O}(\varepsilon)$ term in the non-planar basis diagram was solved by [Kazakov (1984)].

Chapter 9

HQET propagator diagrams

9.1 Crash course of HQET

Heavy Quark Effective Theory (HQET, see, e.g., [Neubert (1994); Manohar and Wise (2000); Grozin (2004)]) is an effective field theory constructed to approximate results of QCD for certain problems with a single heavy quark. Let the heavy quark stay approximately at rest (in some frame); its momentum and energy are $|\vec{p}| \lesssim E$, $|p_0 - m| \lesssim E$. Momenta and energies of light quarks and gluons are $|\vec{k}_i| \lesssim E$, $|k_{0i}| \lesssim E$. Here $E \ll m$ is some fixed characteristic momentum, and we consider the limit $m \to \infty$. Scattering amplitudes and on-shell matrix elements of operators in QCD, expanded up to some order in E/m, can be reproduced from a simpler theory — HQET. The lowest-energy state of this theory ("vacuum") is a single heavy quark at rest, and has energy m. Therefore, for any system containing this quark it is natural to measure energy relative this zero level, i.e., to consider its residual energy $\tilde{p}_0 = p_0 - m$. The free heavy quark has dispersion law (dependence of energy on momentum, or mass shell) $\tilde{p}_0 = \sqrt{m^2 + \vec{p}^2} - m$. Neglecting $1/m$ corrections, we may simplify it to $\tilde{p}_0(\vec{p}) = 0$: the heavy-quark energy is zero, and does not depend on its momentum. The heavy quark at rest is described by a two-component spinor field Q, or a four-component spinor having only upper components: $\gamma_0 Q = Q$. The dispersion law $\tilde{p}_0(\vec{p}) = 0$ follows from the Lagrangian $L = Q^+ i \partial_0 Q$. Reintroducing the interaction with the gluon field by the requirement of gauge invariance, we obtain the leading-order (in $1/m$) HQET Lagrangian

$$L = Q^+ i D_0 Q + \cdots \tag{9.1}$$

where all light-field parts (denoted by dots) are exactly the same as in QCD. The heavy quark interacts with A_0: creates the coulomb chromoelectric field, and reacts to external chromoelectric fields.

The Lagrangian (9.1) gives the heavy quark propagator

$$S(\widetilde{p}) = \frac{1}{\widetilde{p}_0 + i0}, \quad S(x) = -i\theta(x_0)\delta(\vec{x}). \tag{9.2}$$

In the momentum space it depends only on $\widetilde{p}_0$ but not on $\vec{p}$, because we have neglected the kinetic energy. Therefore, in the coordinate space the heavy quark does not move. The vertex is $igv^\mu t^a$, where $v^\mu = (1,0,0,0)$ is the heavy-quark 4-velocity. Heavy-quark loops vanish, because it propagates only forward in time.

HQET is not Lorentz-invariant, because it has a preferred frame — the heavy-quark rest frame. However, it can be rewritten in covariant notations. Momentum of any system containing the heavy quark is decomposed as

$$p = mv + \widetilde{p}, \tag{9.3}$$

where the residual momentum is small: $|\widetilde{p}^\mu| \lesssim E$. The heavy-quark field obeys $\slashed{v}Q_v = Q_v$. The Lagrangian is

$$L = \bar{Q}_v iv \cdot DQ_v + \cdots \tag{9.4}$$

It gives the propagator

$$S(\widetilde{p}) = \frac{1 + \slashed{v}}{2} \frac{1}{\widetilde{p} \cdot v + i0} \tag{9.5}$$

and the vertex $igv^\mu t^a$.

The QCD heavy-quark propagator at large m becomes

$$S(p) = \frac{\slashed{p} + m}{p^2 - m^2} = \frac{m(1 + \slashed{v}) + \widetilde{\slashed{p}}}{2m\widetilde{p} \cdot v + \widetilde{p}^2} = \frac{1 + \slashed{v}}{2} \frac{1}{\widetilde{p} \cdot v} + \mathcal{O}\left(\frac{\widetilde{p}}{m}\right). \tag{9.6}$$

A vertex $ig\gamma^\mu t^a$ sandwiched between two projectors $\frac{1 + \slashed{v}}{2}$ may be replaced by $igv^\mu t^a$. Therefore, at the tree level all QCD diagrams become the corresponding HQET diagrams, up to $1/m$ corrections. In loop diagrams, momenta can be arbitrarily large, and this correspondence breaks down. Renormalization properties of HQET differ from those of QCD. The ultraviolet behaviour of an HQET diagram is determined by the region of loop momenta much larger than the characteristic momentum scale of the process E, but much less than the heavy quark mass m (which tends to infinity from the very beginning). It has nothing to do with the ultraviolet behaviour of the corresponding QCD diagram, which is determined by the region of loop momenta much larger than m.

9.2 One loop

The one-loop HQET propagator diagram (Fig. 9.1) is

$$\int \frac{d^d k}{D_1^{n_1} D_2^{n_2}} = i\pi^{d/2}(-2\omega)^{d-2n_2} I(n_1, n_2)\,,$$

$$D_1 = \frac{k_0 + \omega}{\omega}\,, \quad D_2 = -k^2\,. \tag{9.7}$$

It vanishes if $n_1 \le 0$ or $n_2 \le 0$.

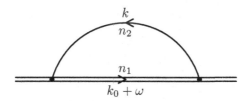

Fig. 9.1   One-loop HQET propagator diagram

The Fourier transform of the one-dimensional (HQET) propagator is

$$\int_{-\infty}^{+\infty} \frac{e^{-i\omega t}}{(-\omega - i0)^n} \frac{d\omega}{2\pi} = \frac{i^n}{\Gamma(n)} t^{n-1} \theta(t)\,, \tag{9.8}$$

$$\int_0^{\infty} e^{i\omega t} t^n \, dt = \frac{(-i)^{n+1}\Gamma(n+1)}{(-\omega - i0)^{n+1}}\,. \tag{9.9}$$

Our diagram in x-space is the product of the heavy propagator (9.8) with the power n_1 and the light propagator (8.3) with the power n_2. The heavy quark stays at rest: $\vec{x} = 0$. Therefore, $-x^2 = (it)^2$ in (8.3)[1]. Our diagram in the x-space is

$$-2^{-2n_2}\pi^{-d/2}\frac{\Gamma(d/2 - n_2)}{\Gamma(n_1)\Gamma(n_2)}(it)^{n_1+2n_2-d-1}\theta(t)\,.$$

Transforming it back to p-space (9.9), we arrive at

$$I(n_1, n_2) = \frac{\Gamma(-d + n_1 + 2n_2)\Gamma(d/2 - n_2)}{\Gamma(n_1)\Gamma(n_2)}\,. \tag{9.10}$$

---

[1]Why not $(-it)^2$? The Wick rotation to the Euclidean x-space is $t = -it_E$. There are no imaginary parts at $t_E > 0$.

9.3 Two loops

There are two generic topologies of two-loop HQET propagator diagrams. The first one is (Fig. 9.2):

$$\int \frac{d^d k_1 \, d^d k_2}{D_1^{n_1} D_2^{n_2} D_3^{n_3} D_4^{n_4} D_5^{n_5}} = -\pi^d (-2\omega)^{2(d-n_3-n_4-n_5)} I(n_1, n_2, n_3, n_4, n_5),$$

$$D_1 = \frac{k_{10} + \omega}{\omega}, \quad D_2 = \frac{k_{20} + \omega}{\omega},$$

$$D_3 = -k_1^2, \quad D_4 = -k_2^2, \quad D_5 = -(k_1 - k_2)^2. \tag{9.11}$$

This diagram is symmetric with respect to $(1 \leftrightarrow 2, 3 \leftrightarrow 4)$. It vanishes if indices of any two adjacent lines are non-positive.

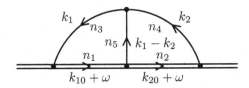

Fig. 9.2 Two-loop HQET propagator diagram I

If $n_5 = 0$, our diagram is just the product of two one-loop diagrams (Fig. 9.3):

$$I(n_1, n_2, n_3, n_4, 0) = I(n_1, n_3) I(n_2, n_4). \tag{9.12}$$

If $n_1 = 0$, the inner loop gives $G(n_3, n_5)/(-k_2^2)^{n_3+n_5-d/2}$, and (Fig. 9.4)

$$I(0, n_2, n_3, n_4, n_5) = G(n_3, n_5) I(n_2, n_4 + n_3 + n_5 - d/2). \tag{9.13}$$

If $n_3 = 0$, the inner loop gives $I(n_1, n_5)/(-k_0)^{n_1+2n_5-d}$, and (Fig. 9.5)

$$I(n_1, n_2, 0, n_4, n_5) = I(n_1, n_5) I(n_2 + n_1 + 2n_5 - d, n_4). \tag{9.14}$$

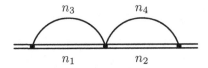

Fig. 9.3 Trivial case $n_5 = 0$

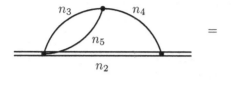

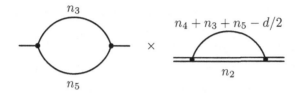

Fig. 9.4 Trivial case $n_1 = 0$

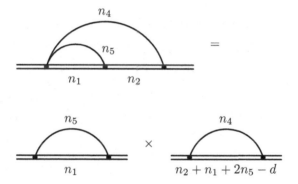

Fig. 9.5 Trivial case $n_3 = 0$

The cases $n_2 = 0$, $n_4 = 0$ are symmetric.

When all $n_i > 0$, we shall use integration by parts [Broadhurst and Grozin (1991)]. The differential operator $\partial/\partial k_2$ applied to the integrand of (9.11) acts as

$$\frac{\partial}{\partial k_2} \rightarrow -\frac{n_2}{D_2}\frac{v}{\omega} + \frac{n_4}{D_4}2k_2 + \frac{n_5}{D_5}2(k_2 - k_1). \tag{9.15}$$

Applying $(\partial/\partial k_2) \cdot k_2$, $(\partial/\partial k_2) \cdot (k_2 - k_1)$ to the integrand of (9.11), we get vanishing integrals. On the other hand, using $k_2 v/\omega = D_2 - 1$, $2(k_2 - k_1) \cdot k_2 = D_3 - D_4 - D_5$, we see that these differential operators are equivalent to inserting

$$d - n_2 - n_5 - 2n_4 + \frac{n_2}{D_2} + \frac{n_5}{D_5}(D_3 - D_4)\,,$$

$$d - n_2 - n_4 - 2n_5 + \frac{n_2}{D_2}D_1 + \frac{n_4}{D_4}(D_3 - D_5)$$

under the integral sign. We obtain the recurrence relations

$$\cdot \quad \left[d - n_2 - n_5 - 2n_4 + n_2\mathbf{2}^+ + n_5\mathbf{5}^+(\mathbf{3}^- - \mathbf{4}^-)\right]I = 0\,, \qquad (9.16)$$

$$\left[d - n_2 - n_4 - 2n_5 + n_2\mathbf{2}^+\mathbf{1}^- + n_4\mathbf{4}^+(\mathbf{3}^- - \mathbf{5}^-)\right]I = 0 \qquad (9.17)$$

(two more relations are obtained by $(1 \leftrightarrow 3, 2 \leftrightarrow 4)$). Similarly, applying the differential operator $(\partial/\partial k_2) \cdot v$ is equivalent to inserting

$$-2\frac{n_2}{D_2} + \frac{n_4}{D_4}4\omega^2(D_2 - 1) + \frac{n_5}{D_5}4\omega^2(D_2 - D_1)\,,$$

and we obtain

$$\left[-2n_2\mathbf{2}^+ + n_4\mathbf{4}^+(\mathbf{2}^- - 1) + n_5\mathbf{5}^+(\mathbf{2}^- - \mathbf{1}^-)\right]I = 0\,. \qquad (9.18)$$

(there is also the symmetric relation, of course).

We can obtain a relation from homogeneity of the integral (9.11) in ω. Applying the operator $\omega \cdot d/d\omega$ to $\omega^{-n_1-n_2}I$ (9.11), we see that it is equivalent to the factor $2(d - n_3 - n_4 - n_5) - n_1 - n_2$. On the other hand, explicit differentiation of $(-\omega D_1)^{-n_1}(-\omega D_2)^{-n_2}$ gives $-n_1/D_1 - n_2/D_2$. Therefore,

$$\left[2(d - n_3 - n_4 - n_5) - n_1 - n_2 + n_1\mathbf{1}^+ + n_2\mathbf{2}^+\right]I = 0\,. \qquad (9.19)$$

This is nothing but the sum of the $(\partial/\partial k_2) \cdot k_2$ relation (9.16) and its mirror-symmetric $(\partial/\partial k_1) \cdot k_1$ relation. A useful relation can be obtained by subtracting $\mathbf{1}^-$ shifted (9.19) from (9.17):

$$\left[d - n_1 - n_2 - n_4 - 2n_5 + 1 - \left(2(d - n_3 - n_4 - n_5) - n_1 - n_2 + 1\right)\mathbf{1}^-\right.$$
$$\left. + n_4\mathbf{4}^+(\mathbf{3}^- - \mathbf{5}^-)\right]I = 0\,. \qquad (9.20)$$

Expressing $I(n_1, n_2, n_3, n_4, n_5)$ from (9.17)

$$I(n_1, n_2, n_3, n_4, n_5) = -\frac{n_2\mathbf{2}^+\mathbf{1}^- + n_4\mathbf{4}^+(\mathbf{3}^- - \mathbf{5}^-)}{d - n_2 - n_4 - 2n_5}I\,, \qquad (9.21)$$

we see that the sum $n_1 + n_3 + n_5$ reduces by 1 (Fig. 8.8). Therefore, applying (9.21) sufficiently many times, we can reduce an arbitrary I integral with integer indices to a combination of integrals with $n_5 = 0$ (Fig. 9.3,

(9.12)), $n_1 = 0$ (Fig. 9.4, (9.13)), $n_3 = 0$ (Fig. 9.5, (9.14)). Of course, if $\max(n_2, n_4) < \max(n_1, n_3)$, it may be more efficient to use the relation mirror-symmetric to (9.17). The relation (9.20) also can be used instead of (9.17).

The second topology of two-loop HQET propagator diagrams is (Fig. 9.6):

$$\int \frac{d^d k_1 \, d^d k_2}{D_1^{n_1} D_2^{n_2} D_3^{n_3} D_4^{n_4} D_5^{n_5}} = -\pi^d (-2\omega)^{2(d-n_4-n_5)} J(n_1, n_2, n_3, n_4, n_5),$$

$$D_1 = \frac{k_{10} + \omega}{\omega}, \quad D_2 = \frac{k_{20} + \omega}{\omega}, \quad D_3 = \frac{(k_1 + k_2)_0 + \omega}{\omega},$$

$$D_4 = -k_1^2, \quad D_5 = -k_2^2. \tag{9.22}$$

This diagram is symmetric with respect to $(1 \leftrightarrow 2, 4 \leftrightarrow 5)$. It vanishes if $n_4 \leq 0$, or $n_5 \leq 0$, or two adjacent heavy indices are non-positive. If $n_3 = 0$, our diagram is just the product of two one-loop diagrams (Fig. 9.3). If $n_1 = 0$, it is also trivial (Fig. 9.5).

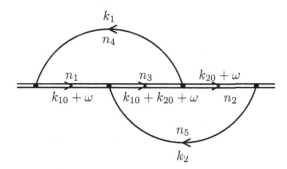

Fig. 9.6 Two-loop HQET propagator diagram J

The denominators in (9.22) are linearly dependent[2]: $D_1 + D_2 - D_3 = 1$. Therefore [Broadhurst and Grozin (1991)]

$$J = (\mathbf{1}^- + \mathbf{2}^- - \mathbf{3}^-)J. \tag{9.23}$$

Applying this relation sufficiently many times, we can kill one of the lines 1, 2, 3, and thus reduce any integral J with integer indices to trivial cases. In fact, we have not enough independent denominators to express all scalar

---

[2]HQET denominators are linear in momenta; there are only two loop momenta, and three denominators cannot be independent.

products in the numerator. Therefore, we have to consider a more general integral than (9.22), containing powers of $k_1 \cdot k_2$ in the numerator. This wider class of integrals can also be reduced to the same trivial cases [Grozin (2000)].

Let's summarize. There are two generic topologies of two-loop HQET propagator diagrams (Fig. 9.7). The integrals with all integer n_i can be expressed as linear combinations of two basis integrals (Fig. 9.8); coefficients are rational functions of d. Here I_n are n-loop HQET sunset diagrams (Fig. 9.9):

$$I_n = \frac{\Gamma(1 + 2n\varepsilon)\Gamma^n(1 - \varepsilon)}{(1 - n(d - 2))_{2n}} . \tag{9.24}$$

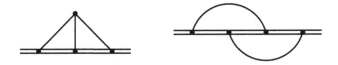

Fig. 9.7   Topologies of two-loop HQET propagator diagram

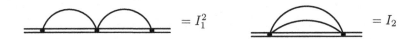

Fig. 9.8   Basis diagrams (all indices equal 1)

Fig. 9.9 n-loop HQET sunset diagrams

9.4 Three loops

There are 10 generic topologies of three-loop HQET propagator diagrams (Fig. 9.10). Diagrams in the first two rows of the figure have one scalar product which cannot be expressed via denominators; those in the third row have one linear relation among heavy denominators, and hence two independent scalar products in the numerator; those in the last row have two relations among heavy denominators, and hence three independent scalar products in the numerator. Integration-by-parts recurrence relations for these diagrams have been investigated by [Grozin (2000)]. They can be used to reduce all integrals of Fig. 9.10, with arbitrary integer powers of denominators and arbitrary numerators, to linear combinations of 8 basis integrals (Fig. 9.11). This algorithm has been implemented in the REDUCE package

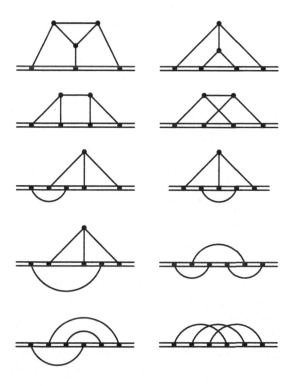

Fig. 9.10   Topologies of three-loop HQET propagator diagrams

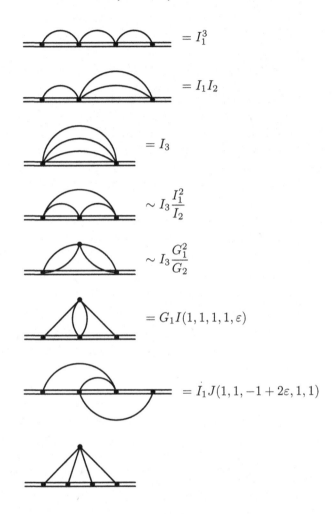

Fig. 9.11   Basis diagrams (all indices equal 1, no numerators)

Grinder [Grozin (2000)][3]. Recently, the heavy-quark propagator and the heavy-light current anomalous dimension have been calculated [Chetyrkin and Grozin (2003)] at 3 loops, using this package.

The first 5 basis integrals are trivial: they are expressed via I_n (9.24) (Fig. 9.9) and G_n (8.18) (Fig. 8.11), and hence via Γ-functions. The integrals 4 and 5 differ from the previous one by replacing a two-loop subdia-

---

[3]The hep-ph version of this paper contains some corrections as compared to the journal version.

gram. In the first case, the two-loop HQET subdiagram (Fig. 9.7) is substituted: $I_2/(-\omega)^{5-2d} \to I_1^2/(-\omega)^{6-2d}$. In the second case, the two-loop massless subdiagram (Fig. 8.9) is substituted: $G_2/(-k^2)^{3-d} \to G_1^2/(-k^2)^{4-d}$. Therefore, these diagrams are obtained from I_3 by multiplying by[4]

$$\frac{I_1^2 I(6-2d,1)}{I_2 I(5-2d,1)} = \frac{3d-7}{2d-5}\frac{I_1^2}{I_2},$$

$$\frac{G_1^2 I(1,4-d)}{G_2 I(1,3-d)} = -2\frac{3d-7}{d-3}\frac{G_1^2}{G_2}.$$

Two basis integrals are proportional to $I(1,1,1,1,n)$ (Sect. 9.6) and $J(1,1,n,1,1)$ (Sect. 9.5), for non-integer n. The last one is truly three-loop and most difficult; it will be discussed in Sect. 10.4.

9.5 $J(1,1,n,1,1)$

This diagram (Fig. 9.12) has been calculated in [Grozin (2000)]. In x-space, it is is

$$\int_{0<t_1<t_2<t} dt_1\, dt_2\, (i(t_2-t_1))^{n-1}(it_2)^{2-d}(i(t-t_1))^{2-d},$$

or, going to Euclidean space $(t_i = -it_{Ei})$,

$$\int_{0<t_1<t_2<t} dt_1\, dt_2\, (t_2-t_1)^{n-1}t_2^{2-d}(t-t_1)^{2-d} = Jt^{n-2d+5},$$

where the power of t is obvious from counting dimensions. Collecting factors from Fourier transforms, we have

$$J(1,1,n,1,1) = \frac{\Gamma(n-2d+6)\Gamma^2(d/2-1)}{\Gamma(n)}J, \qquad (9.25)$$

where the dimensionless integral J is

$$J = \int_{0<t_1<t_2<1} dt_1\, dt_2\, (t_2-t_1)^{n-1}t_2^{2-d}(1-t_1)^{2-d}.$$

The substitution $t_1 = xt_2$ gives

$$J = \int_0^1 dt\, t^{n-d+2} \int_0^1 dx\, (1-x)^{n-1}(1-xt)^{2-d}.$$

---

[4]Grinder uses $B_4 = I_3 I_1^2/I_2$ and $B_5 = I_3 G_1^2/G_2$ as elements of its basis, not the diagrams in Fig. 9.11.

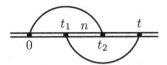

Fig. 9.12 $J(1,1,n,1,1)$ in x-space

Integrations in t and x would factor, if not the factor $(1 - xt)^{2-d}$. Let's expand this troublesome factor according to the Newton binomial:

$$(1 - xt)^{2-d} = \sum_{k=0}^{\infty} \frac{(d-2)_k}{k!} (xt)^k .$$

Here the Pochhammer symbol is

$$(x)_k = \prod_{i=0}^{k-1} (x+i) = \frac{\Gamma(x+k)}{\Gamma(x)} . \tag{9.26}$$

Then the integrations become trivial:

$$\begin{aligned}
J &= \Gamma(n) \sum_{k=0}^{\infty} \frac{(d-2)_k}{(n-d+k+3)\Gamma(n+k+1)} \\
&= \frac{1}{n(n-d+3)} \sum_{k=0}^{\infty} \frac{(d-2)_k(n-d+3)_k}{(n+1)_k(n-d+4)_k} .
\end{aligned} \tag{9.27}$$

Recalling the definition of the hypergeometric function

$${}_3F_2 \left(\begin{matrix} a_1, a_2, a_3 \\ b_1, b_2 \end{matrix} \middle| x \right) = \sum_{k=0}^{\infty} \frac{(a_1)_k(a_2)_k(a_3)_k}{(b_1)_k(b_2)_k} \frac{x^k}{k!} \tag{9.28}$$

and taking into account $(1)_k = k!$, we can rewrite the result as

$$J = \frac{1}{n(n-d+3)} {}_3F_2 \left(\begin{matrix} 1, d-2, n-d+3 \\ n+1, n-d+4 \end{matrix} \middle| 1 \right) ,$$

and hence (9.25)

$$J(1,1,n,1,1) = \frac{\Gamma(n-2d+6)\Gamma^2(d/2-1)}{(n-d+3)\Gamma(n+1)} {}_3F_2 \left(\begin{matrix} 1, d-2, n-d+3 \\ n+1, n-d+4 \end{matrix} \middle| 1 \right) . \tag{9.29}$$

This is the easiest diagram calculation in the world having ${}_3F_2$ as its result: just a double integral.

It is not more difficult to obtain the general result[5]

$$J(n_1, n_2, n_3, n_4, n_5) = \Gamma(n_1 + n_2 + n_3 + 2(n_4 + n_5 - d))$$
$$\times \frac{\Gamma(n_1 + n_3 + 2n_4 - d)\Gamma(d/2 - n_4)\Gamma(d/2 - n_5)}{\Gamma(n_1 + n_2 + n_3 + 2n_4 - d)\Gamma(n_4)\Gamma(n_5)\Gamma(n_1 + n_3)} \qquad (9.30)$$
$$\times {}_3F_2 \left(\begin{matrix} n_1, d - 2n_5, n_1 + n_3 + 2n_4 - d \\ n_1 + n_3, n_1 + n_2 + n_3 + 2n_4 - d \end{matrix} \middle| 1 \right).$$

In order to calculate the two-loop diagram of Fig. 9.6 with a one-loop insertion in the middle heavy line, we need $J(1, 1, n + 2\varepsilon, 1, 1)$. It is easy to shift n_3 by ± 1 using (9.23). Therefore, it is sufficient to find it just for one n. The simplest choice for which the algorithm of ε-expansion (Chap. 11) works is $n = 2$. We obtain [Chetyrkin and Grozin (2003)] from (9.29)

$$J(1, 1, 2 + 2\varepsilon, 1, 1) = \frac{1}{3(d-4)(d-5)(d-6)(2d-9)}$$
$$\times \frac{\Gamma(1 + 6\varepsilon)\Gamma^2(1 - \varepsilon)}{\Gamma(1 + 2\varepsilon)} {}_3F_2 \left(\begin{matrix} 1, 2 - 2\varepsilon, 1 + 4\varepsilon \\ 3 + 2\varepsilon, 2 + 4\varepsilon \end{matrix} \middle| 1 \right). \qquad (9.31)$$

Expansion of this ${}_3F_2$ in ε is (Chap. 11)

$${}_3F_2 \left(\begin{matrix} 1, 2 - 2\varepsilon, 1 + 4\varepsilon \\ 3 + 2\varepsilon, 2 + 4\varepsilon \end{matrix} \middle| 1 \right) = 2 + 6(-2\zeta_2 + 3)\varepsilon$$
$$+ 12(10\zeta_3 - 11\zeta_2 + 6)\varepsilon^2 + 24(-28\zeta_4 + 55\zeta_3 - 27\zeta_2 + 9)\varepsilon^3 \qquad (9.32)$$
$$+ 48(94\zeta_5 - 16\zeta_2\zeta_3 - 154\zeta_4 + 135\zeta_3 - 45\zeta_2 + 12)\varepsilon^4 + \cdots$$

It is not difficult to find several additional terms.

9.6 $I(1, 1, 1, 1, n)$

This diagram (Fig. 9.13) has been calculated by [Beneke and Braun (1994)], using Gegenbauer polynomials in x-space [Chetyrkin *et al.* (1980)]. This is, probably, the simplest example of applying this useful technique, therefore, we shall consider some details of this calculation. In Euclidean x-space,

---

[5]I have presented only the result for $n_1 = n_2 = 1$ in [Grozin (2000)], I have no idea why: the general case is not more difficult.

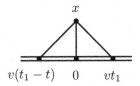

Fig. 9.13 $I(1,1,1,1,n)$ in x-space

this diagram is

$$\int_0^t dt_1 \int d^d x \, \frac{1}{(x^2)^{d/2-n} \left[(x-vt_1)^2\right]^\lambda \left[(x+v(t-t_1))^2\right]^\lambda}$$
$$= \frac{2\pi^{d/2}}{\Gamma(d/2)} I t^{2(n-d)+5} \,,$$

where $\lambda = d/2 - 1$, and the power of t is obvious from counting dimensions. The middle vertex has been chosen as the origin, because the propagator with the non-standard power begins at this vertex (it is much easier to handle propagators beginning at the origin). Collecting factors from Fourier transforms, we have

$$I(1,1,1,1,n) = \frac{2}{\pi} \frac{\Gamma(2(n-d+3))\Gamma(d/2-n)\Gamma^2(d/2-1)}{\Gamma(d/2)\Gamma(n)} I \,. \qquad (9.33)$$

The dimensionless integral I is

$$I = \int_0^1 dt \int_0^\infty dx \, d\hat{x} \, \frac{x^{2n-1}}{\left[(x-vt)^2\right]^\lambda \left[(x+v(1-t))^2\right]^\lambda} \,,$$

where

$$d^d x = \frac{2\pi^{d/2}}{\Gamma(d/2)} x^{d-1} dx \, d\hat{x} \,,$$

and the angular integration measure $d\hat{x}$ is normalized:

$$\int d\hat{x} = 1 \,.$$

This problem is more difficult than the one in Sect. 9.5: now we have d-dimensional integration over the coordinates of the light vertex. It can be simplified by expanding two massless propagators (whose ends are not at the origin) in series in Gegenbauer polynomials [Chetyrkin *et al.* (1980)]

$$\frac{1}{\left[(x-y)^2\right]^\lambda} = \frac{1}{\max^{2\lambda}(x,y)} \sum_{k=0}^\infty T^k(x,y) C_k^\lambda(\hat{x} \cdot \hat{y}) \,, \qquad (9.34)$$

where

$$T(x, y) = \min\left(\frac{x}{y}, \frac{y}{x}\right).$$

Then the angular integration can be easily done using the orthogonality relation

$$\int d\hat{x}\, C_{k_1}^\lambda(\hat{a} \cdot \hat{x}) C_{k_2}^\lambda(\hat{b} \cdot \hat{x}) = \delta_{k_1 k_2} \frac{\lambda}{\lambda + k_1} C_{k_1}^\lambda(\hat{a} \cdot \hat{b}). \qquad (9.35)$$

In our particular case, the unit vectors $\hat{a}$ and $\hat{b}$ are v and $-v$, and the result involves

$$C_k^\lambda(-1) = (-1)^k C_k^\lambda(1) = (-1)^k \frac{\Gamma(2\lambda + k)}{k!\, \Gamma(2\lambda)}.$$

Now we have a single sum:

$$I = \frac{d-2}{\Gamma(d-2)} \int_0^1 dt \sum_{k=0}^\infty \frac{(-1)^k}{k!} \frac{\Gamma(d+k-2)}{d+2k-2} I_k(t),$$

$$I_k(t) = \int_0^\infty dx \frac{x^{2n-1} [T(x,t) T(x, 1-t)]^k}{[\max(x,t)\max(x, 1-t)]^{d-2}}.$$

The contributions of the regions $t < \frac{1}{2}$ and $t > \frac{1}{2}$ are equal, and we may consider the first of them and double it. The radial integral (in x) has to be calculated in several intervals separately, because of max and T in (9.34):

$$I_k(t) = \int_0^t dx \frac{x^{2n-1} \left[\dfrac{x}{t} \dfrac{x}{1-t}\right]^k}{[t(1-t)]^{d-2}} + \int_t^{1-t} dx \frac{x^{2n-1} \left[\dfrac{t}{x} \dfrac{x}{1-t}\right]^k}{[x(1-t)]^{d-2}}$$

$$+ \int_{1-t}^\infty dx \frac{x^{2n-1} \left[\dfrac{t}{x} \dfrac{1-t}{x}\right]^k}{[x^2]^{d-2}}.$$

The result is

$$I = \frac{d-2}{(d-2n-2)\Gamma(d-2)} \int_0^{1/2} dt \sum_{k=0}^\infty \frac{(-1)^k}{k!} \Gamma(d+k-2)$$

$$\times \left[\frac{t^{-d+2n+k+2}(1-t)^{-d-k+2}}{n+k} - \frac{t^k(1-t)^{-2d+2n-k+4}}{d-n+k-2}\right].$$

Unfortunately, the integral in t cannot be immediately evaluated. [Beneke and Braun (1994)] had to do some more juggling to obtain

$$
\begin{aligned}
I(1,1,1,1,n) = {} & 2\Gamma\left(\frac{d}{2}-1\right)\Gamma\left(\frac{d}{2}-n-1\right) \\
& \times \left[\frac{\Gamma(2n-2d+6)}{(2n-d+3)\Gamma(n+1)} \, {}_3F_2\left(\begin{array}{c}1,d-2,2n-d+3\\n+1,2n-d+4\end{array}\bigg|\,1\right)\right. \\
& \left. -\frac{\Gamma(d-n-2)\Gamma^2(n-d+3)}{\Gamma(d-2)}\right].
\end{aligned}
\tag{9.36}
$$

This result can also be rewritten [Beneke and Braun (1994)] as

$$
\begin{aligned}
I(1,1,1,1,n) = {} & \frac{\Gamma\left(\frac{d}{2}-1\right)\Gamma\left(\frac{d}{2}-n-1\right)}{\Gamma(d-2)}\left[2\frac{\Gamma(2n-d+3)\Gamma(2n-2d+6)}{(n-d+3)\Gamma(3n-2d+6)}\times\right. \\
& {}_3F_2\left(\begin{array}{c}n-d+3,n-d+3,2n-2d+6\\n-d+4,3n-2d+6\end{array}\bigg|\,1\right) \left.-\Gamma(d-n-2)\Gamma^2(n-d+3)\right],
\end{aligned}
\tag{9.37}
$$

this series has a larger region of convergence.

In order to calculate the two-loop diagram of Fig. 9.2 with a one-loop insertion in the middle light line, we need $I(1,1,1,1,n+\varepsilon)$. It is easy to shift n_5 by ± 1 using the relation [Grozin (2000)]

$$
\begin{aligned}
& \left[(d-2n_5-4)\mathbf{5}^+ - 2(d-n_5-3)\right]I(1,1,1,1,n_5) = \\
& \left[(2d-2n_5-7)\mathbf{1}^-\mathbf{5}^+ - \mathbf{3}^-\mathbf{4}^+\mathbf{5}^- + \mathbf{1}^-\mathbf{3}^+\right]I(1,1,1,1,n_5),
\end{aligned}
\tag{9.38}
$$

which follows from integration-by-parts relations (Sect. 9.3) (all terms in its right-hand side are trivial). Therefore, it is sufficient to find it just for one n. The simplest choice for which the algorithm of ε-expansion (Chap. 11) works is to use $n=1$ in (9.37). We obtain [Chetyrkin and Grozin (2003)]

$$
I(1,1,1,1,1+\varepsilon) = \frac{4\Gamma(1-\varepsilon)}{9(d-3)(d-4)^2}\times
\tag{9.39}
$$

$$
\left[\frac{\Gamma(1+4\varepsilon)\Gamma(1+6\varepsilon)}{\Gamma(1+7\varepsilon)}\,{}_3F_2\left(\begin{array}{c}3\varepsilon,3\varepsilon,6\varepsilon\\1+3\varepsilon,1+7\varepsilon\end{array}\bigg|\,1\right)-\Gamma^2(1+3\varepsilon)\Gamma(1-3\varepsilon)\right].
$$

This ${}_3F_2$ function has three upper indices $\sim \varepsilon$; therefore, its expansion (apart from the leading 1) starts from $\mathcal{O}(\varepsilon^3)$:

$$
{}_3F_2 \left(\begin{array}{c} 3\varepsilon, 3\varepsilon, 6\varepsilon \\ 1 + 3\varepsilon, 1 + 7\varepsilon \end{array} \middle| \, 1 \right) = 1 + 54\zeta_3\varepsilon^3 - 513\zeta_4\varepsilon^4
$$
$$
+ \, 54(25\zeta_5 + 28\zeta_2\zeta_3)\varepsilon^5 + \cdots
$$

(9.40)

It is not difficult to find several additional terms.

Chapter 10

Massive on-shell propagator diagrams

In this Chapter, we shall consider self-energy diagrams of a massive particle with mass m, when the external momentum is on the mass shell: $p^2 = m^2$, or $p = mv$. Why is this special case interesting? First of all, in order to calculate any S-matrix elements, one has to consider mass [Gray *et al.* (1990); Melnikov and van Ritbergen (2000b)] and wave-function [Broadhurst *et al.* (1991); Melnikov and van Ritbergen (2000c)] renormalization in the on-shell scheme. Calculations of form factors and their derivatives at the point $q = 0$ belong to this class, e.g., the anomalous magnetic moment [Laporta and Remiddi (1996)] and electron charge radius [Melnikov and van Ritbergen (2000b)] in QED. Finally, such calculations are used for QCD/HQET matching [Broadhurst and Grozin (1995); Czarnecki and Grozin (1997)].

10.1 One loop

The one-loop on-shell propagator diagram (Fig. 10.1) is

$$
\int \frac{d^d k}{D_1^{n_1} D_2^{n_2}} = i\pi^{d/2} m^{d-2(n_1+n_2)} M(n_1, n_2) \, ,
$$

$$
D_1 = m^2 - (k + mv)^2 \, , \quad D_2 = -k^2 \, .
$$

(10.1)

It vanishes if $n_1 \leq 0$.

The definition (10.1) of $M(n_1, n_2)$ expressed via the dimensionless Euclidean momentum $K = k_E/m$ becomes

$$
\int \frac{d^d K}{(K^2 - 2iK_0)^{n_1} (K^2)^{n_2}} = \pi^{d/2} M(n_1, n_2) \, .
$$

The definition (9.7) of the one-loop HQET integral $I(n_1, n_2)$ expressed via

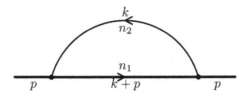

Fig. 10.1   One-loop on-shell propagator diagram

the dimensionless Euclidean momentum $K = k_E/(-2\omega)$ is

$$\int \frac{d^d K}{(1 - 2iK_0)^{n_1}(K^2)^{n_2}} = \pi^{d/2} I(n_1, n_2).$$

Inversion (8.6) transforms the on-shell denominator to the HQET one:

$$K^2 - 2iK_0 = \frac{1 - 2iK_0'}{K'^2} \tag{10.2}$$

(and vice versa). Therefore, the problem of calculating $M(n_1, n_2)$ reduces to the previously solved one for $I(n_1, n_2)$ (Sect. 9.2):

$$M(n_1, n_2) = I(n_1, d - n_1 - n_2) = \frac{\Gamma(d - n_1 - 2n_2)\Gamma(-d/2 + n_1 + n_2)}{\Gamma(n_1)\Gamma(d - n_1 - n_2)} \tag{10.3}$$

(Fig. 10.2).

Fig. 10.2   Inversion relation

10.2 Two loops

There are two generic topologies of two-loop on-shell propagator diagrams. The first one is (Fig. 10.3):

$$\int \frac{d^d k_1 \, d^d k_2}{D_1^{n_1} D_2^{n_2} D_3^{n_3} D_4^{n_4} D_5^{n_5}} = -\pi^d m^{2(d - \sum n_i)} M(n_1, n_2, n_3, n_4, n_5) \,,$$
$$D_1 = m^2 - (k_1 + mv)^2 \,, \quad D_2 = m^2 - (k_2 + mv)^2 \,,$$
$$D_3 = -k_1^2 \,, \quad D_4 = -k_2^2 \,, \quad D_5 = -(k_1 - k_2)^2 \,.$$
(10.4)

The second topology is (Fig. 10.4):

$$\int \frac{d^d k_1 \, d^d k_2}{D_1^{n_1} D_2^{n_2} D_3^{n_3} D_4^{n_4} D_5^{n_5}} = -\pi^d m^{2(d - \sum n_i)} N(n_1, n_2, n_3, n_4, n_5) \,,$$
$$D_1 = m^2 - (k_1 + mv)^2 \,, \quad D_2 = m^2 - (k_2 + mv)^2 \,,$$
$$D_3 = m^2 - (k_1 + k_2 + mv)^2 \,, \quad D_4 = -k_1^2 \,, \quad D_5 = -k_2^2 \,.$$
(10.5)

The integrals of both generic topologies (Fig. 10.5) with all integer n_i can be reduced, by using integration-by-parts recurrence relations, to linear

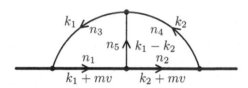

Fig. 10.3   Two-loop on-shell propagator diagram M

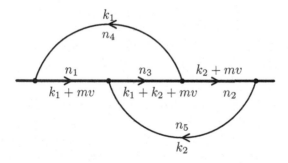

Fig. 10.4   Two-loop on-shell propagator diagram N

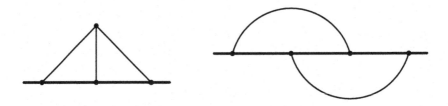

Fig. 10.5   Topologies of two-loop on-shell propagator diagrams

combinations of three basis integrals (Fig. 10.6); coefficients are rational functions of d. Only the topology N involve the last, and the most difficult, basis integral in Fig. 10.6. The reduction algorithm has been constructed and implemented as a REDUCE package RECURSOR in [Gray *et al.* (1990); Broadhurst *et al.* (1991); Broadhurst (1992)]. It also has been implemented as a FORM package SHELL2 [Fleischer and Tarasov (1992); Fleischer and Kalmykov (2000)].

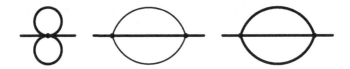

Fig. 10.6   Basis diagrams (all indices equal 1)

The first two basis integrals are trivial. The n-loop on-shell sunset M_n (Fig. 10.7) is

$$M_n =$$

$$\frac{(nd - 4n + 1)_{2(n-1)}}{\left(n + 1 - n\frac{d}{2}\right)_n \left((n+1)\frac{d}{2} - 2n - 1\right)_n \left(n\frac{d}{2} - 2n + 1\right)_{n-1} \left(n - (n-1)\frac{d}{2}\right)_{n-1}}$$

$$\times \frac{\Gamma(1 + (n-1)\varepsilon)\Gamma(1 + n\varepsilon)\Gamma(1 - 2n\varepsilon)\Gamma^n(1 - \varepsilon)}{\Gamma(1 - n\varepsilon)\Gamma(1 - (n+1)\varepsilon)}. \tag{10.6}$$

In the one-loop case, it can be reduced to the one-loop massive vacuum diagram

$$M_1 = -\frac{1}{2}\frac{d - 2}{d - 3}V_1, \quad V_1 = \frac{4\Gamma(1 + \varepsilon)}{(d - 2)(d - 4)} \tag{10.7}$$

(Fig. 10.8).

Fig. 10.7 n-loop on-shell sunset diagrams

$$= -\frac{1}{2}\frac{d-2}{d-3}$$

Fig. 10.8   One-loop on-shell diagram

Instead of using $N(1,1,1,0,0)$ (the last diagram in Fig. 10.6, which is divergent) as an element of the basis, one may use the convergent diagram $N(1,1,1,1,1)$ (the second one in Fig. 10.5). It has been calculated in [Broadhurst (1992)]:

$$
N(1,1,1,1,1) = \frac{\Gamma^2(1+\varepsilon)}{\varepsilon(1-2\varepsilon)}\left[-\frac{4}{1+2\varepsilon}\,{}_3F_2\left(\begin{array}{c}1,\frac{1}{2}-\varepsilon,\frac{1}{2}-\varepsilon\\ \frac{3}{2}+\varepsilon,\frac{3}{2}\end{array}\middle|\,1\right)\right.
$$
$$
+ \frac{1}{4\varepsilon^2}\left(\frac{\Gamma^2(1-\varepsilon)\Gamma(1-4\varepsilon)\Gamma(1+2\varepsilon)}{\Gamma(1-2\varepsilon)\Gamma(1-3\varepsilon)\Gamma(1+\varepsilon)}-1\right)
$$
$$
\left.+ 2^{-6\varepsilon}\frac{\pi^2}{3}\frac{\Gamma(1+2\varepsilon)\Gamma(1+3\varepsilon)}{\Gamma^5(1+\varepsilon)}\right]. \tag{10.8}
$$

Its value at $\varepsilon = 0$ is [Broadhurst (1990)]

$$
N(1,1,1,1,1) = \pi^2\log 2 - \frac{3}{2}\zeta_3 + \mathcal{O}(\varepsilon). \tag{10.9}
$$

[Broadhurst (1992)] discovered a symmetry group of a class of ${}_3F_2$ functions like the one in (10.8); this allowed him to obtain (10.9) and two further terms, up to $\mathcal{O}(\varepsilon^2)$, by purely algebraic methods. Later, he expanded the hypergeometric function in (10.9) in a straightforward way, obtaining multiple sums as coefficients of the series in ε; he has summed these series analytically, up to the ε^5 term, thus obtaining $N(1,1,1,1,1)$ up to ε^4 [Broadhurst (1996)].

Using inversion of Euclidean dimensionless momenta (8.6) and (10.2), [Broadhurst and Grozin (1995a)] obtained the relation of Fig. 10.9.

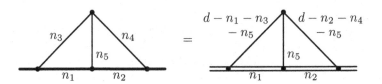

Fig. 10.9   Inversion relation

10.3   Two loops, two masses

In realistic theories, there are several massive particles with different masses, as well as massless particles (QED with e, μ, τ; QCD with c, b, t; the Standard Model). Therefore, diagrams like in Fig. 10.10 appear. Here the thin line is massless, the thick solid line has mass m, and the thick dashed one has mass m'. Combining identical denominators, we obtain the topology shown in Fig. 10.11.

Fig. 10.10   Two-loop on-shell propagator diagram

Fig. 10.11   Topology of two-loop on-shell propagator diagrams with two masses

An algorithm for calculating such diagrams based on integration-by-parts recurrence relations has been constructed by [Davydychev and Grozin (1999)]. The algorithm has been implemented in REDUCE. There is one scalar product which cannot be expressed via the denominators. The integrals of Fig. 10.11 with arbitrary integer powers of the denominators, and with arbitrary non-negative power of the scalar product in the numerator, can be expressed as linear combinations of four basis integrals (Fig. 10.12); coefficients are rational functions of d, m, and m'.

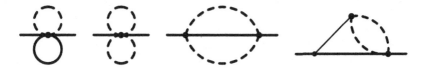

Fig. 10.12   Basis two-loop on-shell propagator diagrams with 2 masses

Instead of using the last diagram in Fig. 10.12 as an element of the basis, one may prefer to use the sunset diagram (the third one in Fig. 10.12) with one of the denominators squared (either the one with mass m or the one with mass m'). Therefore, the reduction algorithm for the sunset diagrams with the masses m, m', m' and $p^2 = m^2$, constructed by [Onishchenko and Veretin (2002)], is just a particular case of the algorithm from [Davydychev and Grozin (1999)].

The first two basis integrals in Fig. 10.12 are just products of one-loop massive vacuum diagrams, and thus trivial. The two non-trivial ones have been calculated [Davydychev and Grozin (1999)] using the Mellin–Barnes representation technique:

$$
\frac{I_0}{\Gamma^2(1+\varepsilon)} = -\frac{(m'^2)^{1-2\varepsilon}}{\varepsilon^2(1-\varepsilon)} \left\{ \frac{1}{1-2\varepsilon} {}_3F_2\left(\begin{matrix} 1, \frac{1}{2}, -1+2\varepsilon \\ 2-\varepsilon, \frac{1}{2}+\varepsilon \end{matrix} \middle| \frac{m^2}{m'^2} \right) \right.
$$
$$
\left. + \left(\frac{m^2}{m'^2} \right)^{1-\varepsilon} {}_3F_2\left(\begin{matrix} 1, \varepsilon, \frac{3}{2}-\varepsilon \\ 3-2\varepsilon, \frac{3}{2} \end{matrix} \middle| \frac{m^2}{m'^2} \right) \right\},
$$
$$
\frac{I_1}{\Gamma^2(1+\varepsilon)} = -\frac{(m'^2)^{-2\varepsilon}}{\varepsilon^2} \left\{ \frac{1}{2(1-\varepsilon)(1+2\varepsilon)} {}_3F_2\left(\begin{matrix} 1, \frac{1}{2}, 2\varepsilon \\ 2-\varepsilon, \frac{3}{2}+\varepsilon \end{matrix} \middle| \frac{m^2}{m'^2} \right) \right.
$$
$$
\left. - \left(\frac{m^2}{m'^2} \right)^{1-\varepsilon} \frac{1}{1-2\varepsilon} {}_3F_2\left(\begin{matrix} 1, \varepsilon, \frac{1}{2}-\varepsilon \\ 2-2\varepsilon, \frac{3}{2} \end{matrix} \middle| \frac{m^2}{m'^2} \right) \right\}. \tag{10.10}
$$

If $m' = m$, they are not independent — I_1 can be reduced to I_0 (the last diagram in Fig. 10.6) using integration by parts (Sect. 10.2):

$$
I_1 = -\frac{3d-8}{4(d-4)} I_0 - \frac{3(d-2)^2}{8(d-3)(d-4)} V_1^2. \tag{10.11}
$$

We can express $N(1,1,1,1,1)$ via I_0 or I_1 (10.10); in the second case, for example, we have

$$
N(1,1,1,1,1) = -\frac{\Gamma^2(1+\varepsilon)}{(d-4)^2} \bigg\{ (3d-10)
$$

$$
\times \left[\frac{1}{(d-2)(d-5)} {}_3F_2\left(\begin{matrix} 1, 2\varepsilon, \frac{1}{2} \\ 2-\varepsilon, \frac{3}{2}+\varepsilon \end{matrix} \middle| 1 \right) + \frac{1}{d-3} {}_3F_2\left(\begin{matrix} 1, \varepsilon, \frac{1}{2}-\varepsilon \\ 2-2\varepsilon, \frac{3}{2} \end{matrix} \middle| 1 \right) \right]
$$

$$
+ \frac{1}{(d-3)(d-4)} \left[2\frac{\Gamma^2(1-\varepsilon)\Gamma(1+2\varepsilon)\Gamma(1-4\varepsilon)}{\Gamma(1+\varepsilon)\Gamma(1-2\varepsilon)\Gamma(1-3\varepsilon)} - 3 \right] \bigg\}. \tag{10.12}
$$

This should be equivalent to (10.8).

The basis integrals (10.10), expanded [Davydychev and Grozin (1999)] in ε up to $\mathcal{O}(1)$,

$$
\frac{I_0}{\Gamma^2(1+\varepsilon)} = -m^{2-4\varepsilon} \left[\frac{1}{2\varepsilon^2} + \frac{5}{4\varepsilon} + 2(1-r^2)^2(L_+ + L_-) - 2\log^2 r + \frac{11}{8} \right]
$$

$$
- m'^{2-4\varepsilon} \left[\frac{1}{\varepsilon^2} + \frac{3}{\varepsilon} - 2\log r + 6 \right] + \mathcal{O}(\varepsilon),
$$

$$
\frac{I_1}{\Gamma^2(1+\varepsilon)} = m^{-4\varepsilon} \left[\frac{1}{2\varepsilon^2} + \frac{5}{2\varepsilon} + 2(1+r)^2 L_+ + 2(1-r)^2 L_- - 2\log^2 r \right.
$$

$$
\left. + \frac{19}{2} \right] + \mathcal{O}(\varepsilon), \tag{10.13}
$$

(where $r = m'/m$), are expressed via dilogarithms:

$$
\begin{aligned}
L_+ &= -\operatorname{Li}_2(-r) + \tfrac{1}{2}\log^2 r - \log r \log(1+r) - \tfrac{1}{6}\pi^2 \\
&= \operatorname{Li}_2(-r^{-1}) + \log r^{-1} \log(1+r^{-1}), \\
L_- &= \operatorname{Li}_2(1-r) + \tfrac{1}{2}\log^2 r + \tfrac{1}{6}\pi^2 \\
&= -\operatorname{Li}_2(1-r^{-1}) + \tfrac{1}{6}\pi^2 \\
&= -\operatorname{Li}_2(r) + \tfrac{1}{2}\log^2 r - \log r \log(1-r) + \tfrac{1}{3}\pi^2 \quad (r<1) \\
&= \operatorname{Li}_2(r^{-1}) + \log r^{-1} \log(1-r^{-1}) \quad (r>1), \\
L_+ + L_- &= \tfrac{1}{2}\operatorname{Li}_2(1-r^2) + \log^2 r + \tfrac{1}{12}\pi^2 \\
&= -\tfrac{1}{2}\operatorname{Li}_2(1-r^{-2}) + \tfrac{1}{12}\pi^2. \tag{10.14}
\end{aligned}
$$

The sunset basis integral I_0 (10.10) (the third one in Fig. 10.12) was recently expanded [Argeri *et al.* (2002)] up to ε^5; the coefficients involve harmonic polylogarithms of r^{-1}. In particular, at $r = 1$, the $\mathcal{O}(\varepsilon^3)$ term in (10.9) can be obtained.

10.4   Three loops

There are 11 generic topologies of three-loop massive on-shell propagator diagrams (Fig. 10.13). Ten of them are the same as in HQET (Fig. 9.10), plus one additional topology with a heavy loop (in HQET, diagrams with heavy-quark loops vanish). Integration-by-parts recurrence relations for these diagrams have been investigated by [Melnikov and van Ritbergen (2000c)]. They can be used to reduce all integrals of Fig. 10.13, with arbitrary integer powers of denominators and arbitrary numerators, to linear combinations of the basis integrals in Fig. 10.14. This algorithm has been implemented in the FORM package SHELL3 [Melnikov and van Ritbergen

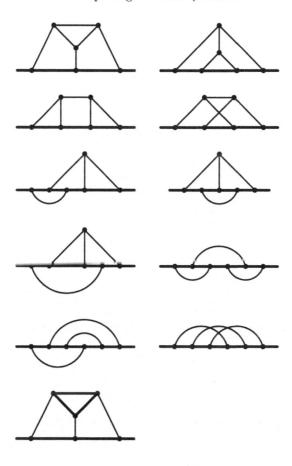

Fig. 10.13   Topologies of three-loop massive on-shell propagator diagrams

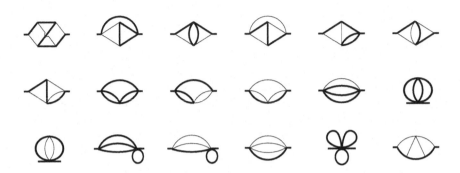

Fig. 10.14   Basis diagrams (all indices equal 1, no numerators)

(2000c)]. Some of the basis integrals are trivial. Others were found, to some orders in ε, during many years, in the course of calculation of the anomalous magnetic moment in QED at three loops [Laporta and Remiddi (1996)]. Some additional terms of ε expansions were obtained by [Melnikov and van Ritbergen (2000c)].

Performing inversion (8.6) of the loop momenta, we obtain the relations in Fig. 10.15. For example, the HQET ladder diagram with all indices $n_i = 1$ is convergent; its value at $d = 4$ is related [Czarnecki and Melnikov (2002)] to a massive on-shell diagram (Fig. 10.16) by the second inversion relation. This is one of the basis diagrams of Fig. 10.14, and its value at $d = 4$ is known (Fig. 10.16). Calculating this ladder diagram with Grinder (Sect. 9.4) and solving for the most difficult HQET three-loop basis integral

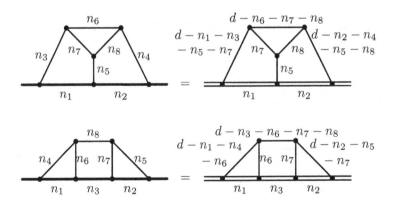

Fig. 10.15   Inversion relations

Fig. 10.16   Inversion relation: all $n_i = 1$, $d = 4$

in Fig. 9.11, we obtain the ε expansion of this integral up to $\mathcal{O}(\varepsilon)$. This concludes the investigation of the basis integrals of Fig. 9.11 (Sects. 9.4–9.6), and allows one to solve three-loop propagator problems in HQET up to terms $\mathcal{O}(1)$ (see [Chetyrkin and Grozin (2003)]).

Chapter 11

Hypergeometric functions and multiple ζ values

In this Section, we shall consider some mathematical methods useful for expanding massless and HQET diagrams in ε.

11.1 Multiple ζ values

As you may have noticed, expansions of various Feynman integrals in ε contain lots of ζ-functions. They appear when expanding Γ-functions:

$$\Gamma(1+\varepsilon) = \exp\left[-\gamma\varepsilon + \sum_{n=2}^{\infty} \frac{(-1)^n \zeta_n}{n} \varepsilon^n\right]. \tag{11.1}$$

They also appear, together with their generalizations — multiple ζ values, when expanding hypergeometric functions of unit argument with indices going to integers at $\varepsilon \to 0$ (Sect. 11.2, 11.3). Such hypergeometric functions appear in massless and HQET diagrams (Sect. 8.5, 9.5, 9.6). In this Section, we shall briefly discuss multiple ζ values; more details can be found in [Borwein *et al.* (2001)], where more general functions are also considered.

The Riemann ζ-function is defined by

$$\zeta_s = \sum_{n>0} \frac{1}{n^s}. \tag{11.2}$$

Let's define multiple ζ values as

$$\zeta_{s_1 s_2} = \sum_{n_1 > n_2 > 0} \frac{1}{n_1^{s_1} n_2^{s_2}},$$

$$\zeta_{s_1 s_2 s_3} = \sum_{n_1 > n_2 > n_3 > 0} \frac{1}{n_1^{s_1} n_2^{s_2} n_3^{s_3}}, \tag{11.3}$$

and so on. These series converge at $s_1 > 1$. The number of summations is called *depth* k; $s = s_1 + \cdots + s_k$ is called *weight*. All identities we shall consider relate terms of the same weight (where the weight of a product is the sum of the factors' weights).

For example, there is 1 convergent sum at weight 2:

$$\zeta_2,$$

2 at weight 3:

$$\zeta_3, \quad \zeta_{21},$$

4 at weight 4:

$$\zeta_4, \quad \zeta_{31}, \quad \zeta_{22}, \quad \zeta_{211},$$

and 8 at weight 5:

$$\zeta_5, \quad \zeta_{41}, \quad \zeta_{32}, \quad \zeta_{23}, \quad \zeta_{311}, \quad \zeta_{221}, \quad \zeta_{212}, \quad \zeta_{2111}.$$

Suppose we want to multiply $\zeta_s \zeta_{s_1 s_2}$:

$$\zeta_s \zeta_{s_1 s_2} = \sum_{\substack{n>0 \\ n_1 > n_2 > 0}} \frac{1}{n^s n_1^{s_1} n_2^{s_2}}.$$

Here n can be anywhere with respect to n_1, n_2. There are 5 contributions:

$$\sum_{n > n_1 > n_2 > 0} \frac{1}{n^s n_1^{s_1} n_2^{s_2}} = \zeta_{s s_1 s_2},$$

$$\sum_{n = n_1 > n_2 > 0} \frac{1}{n^s n_1^{s_1} n_2^{s_2}} = \zeta_{s+s_1, s_2},$$

$$\sum_{n_1 > n > n_2 > 0} \frac{1}{n^s n_1^{s_1} n_2^{s_2}} = \zeta_{s_1 s s_2},$$

$$\sum_{n_1 > n = n_2 > 0} \frac{1}{n^s n_1^{s_1} n_2^{s_2}} = \zeta_{s_1, s+s_2},$$

$$\sum_{n_1 > n_2 > n > 0} \frac{1}{n^s n_1^{s_1} n_2^{s_2}} = \zeta_{s_1 s_2 s}.$$

Therefore,

$$\zeta_s \zeta_{s_1 s_2} = \zeta_{s s_1 s_2} + \zeta_{s+s_1, s_2} + \zeta_{s_1 s s_2} + \zeta_{s_1, s+s_2} + \zeta_{s_1 s_2 s}. \tag{11.4}$$

This process reminds shuffling cards (Fig. 11.1). The order of cards in the upper deck, as well as in the lower one, is kept fixed. We sum over all possible shufflings. Unlike real playing cards, however, two cards may be exactly on top of each other. In this case, they are stuffed together: a single card (which is their sum) appears in the resulting deck. A mathematical jargon term for such shuffling with (possible) stuffing is *stuffling*, see [Borwein *et al.* (2001)].

Fig. 11.1 Stuffling: shuffling with (possible) stuffing

For example, the stuffling relations for weights $s \leq 5$ are

$$\zeta_3^2 = \zeta_{33} + \zeta_4 + \zeta_{33},$$
$$\zeta_2\zeta_3 = \zeta_{23} + \zeta_5 + \zeta_{32},$$
$$\zeta_2\zeta_{21} = \zeta_{221} + \zeta_{41} + \zeta_{221} + \zeta_{23} + \zeta_{212}.$$
(11.5)

Now we are going to derive an integral representation of multiple ζ values. Let's consider the integral

$$\int_0^1 \frac{dx_1}{x_1} \int_0^{x_1} \frac{dx_2}{x_2} \int_0^{x_2} \frac{dx_3}{x_3} \int_0^{x_3} \frac{dx_4}{x_4} x_4^n =$$
$$\int_0^1 \frac{dx_1}{x_1} \int_0^{x_1} \frac{dx_2}{x_2} \int_0^{x_2} \frac{dx_3}{x_3} x_3^n \cdot \frac{1}{n} =$$
$$\int_0^1 \frac{dx_1}{x_1} \int_0^{x_1} \frac{dx_2}{x_2} x_2^n \cdot \frac{1}{n^2} =$$

$$\int_0^1 \frac{dx_1}{x_1} x_1^n \cdot \frac{1}{n^3} = \frac{1}{n^4} \,.$$

It is easy to guess that

$$\frac{1}{n^s} = \int_{1>x_1>\cdots>x_s>0} \frac{dx_1}{x_1} \cdots \frac{dx_s}{x_s} x_s^n \,.$$

But we need the sum (11.2). This is also easy:

$$\zeta_s = \int_{1>x_1>\cdots>x_s>0} \frac{dx_1}{x_1} \cdots \frac{dx_{s-1}}{x_{s-1}} \frac{dx_s}{1-x_s} \,. \tag{11.6}$$

Let's introduce short notation:

$$\omega_0 = \frac{dx}{x} \,, \quad \omega_1 = \frac{dx}{1-x} \,.$$

All integrals will always have the integration region $1 > x_1 > \cdots > x_s > 0$. Then (11.6) becomes

$$\zeta_s = \int \omega_0^{s-1} \omega_1 \,. \tag{11.7}$$

This can be generalized to multiple sums:

$$\begin{aligned}
\zeta_{s_1 s_2} &= \int \omega_0^{s_1-1} \omega_1 \, \omega_0^{s_2-1} \omega_1 \,, \\
\zeta_{s_1 s_2 s_3} &= \int \omega_0^{s_1-1} \omega_1 \, \omega_0^{s_2-1} \omega_1 \, \omega_0^{s_3-1} \omega_1 \,,
\end{aligned} \tag{11.8}$$

and so on.

Suppose we want to multiply $\zeta_2 \zeta_2$:

$$\zeta_2^2 = \int_{1>x_1>x_2>0} \omega_0 \omega_1 \cdot \int_{1>x_1'>x_2'>0} \omega_0 \omega_1 \,.$$

The ordering of primed and non-primed integration variables is not fixed.

There are 6 contributions:

$$1 > x_1 > x_2 > x_1' > x_2' > 0 : \quad \int \omega_0 \omega_1 \omega_0 \omega_1 = \zeta_{22} \,,$$

$$1 > x_1 > x_1' > x_2 > x_2' > 0 : \quad \int \omega_0 \omega_0 \omega_1 \omega_1 = \zeta_{31} \,,$$

$$1 > x_1 > x_1' > x_2' > x_2 > 0 : \quad \int \omega_0 \omega_0 \omega_1 \omega_1 = \zeta_{31} \,,$$

$$1 > x_1' > x_1 > x_2 > x_2' > 0 : \quad \int \omega_0 \omega_0 \omega_1 \omega_1 = \zeta_{31} \,,$$

$$1 > x_1' > x_1 > x_2' > x_2 > 0 : \quad \int \omega_0 \omega_0 \omega_1 \omega_1 = \zeta_{31} \,,$$

$$1 > x_1' > x_2' > x_1 > x_2 > 0 : \quad \int \omega_0 \omega_1 \omega_0 \omega_1 = \zeta_{22} \,.$$

Therefore,

$$\zeta_2^2 = 4\zeta_{31} + 2\zeta_{22} \,. \tag{11.9}$$

Now we are multiplying integrals, not sums. Therefore, our cards are now infinitely thin, and cannot be exactly on top of each other. There are just two kinds of cards: ω_0 and ω_1, and we sum over all possible shufflings of two decks (Fig. 11.2).

Fig. 11.2   Shuffling

For example, the shuffling relations for weight 5 are

$$\zeta_2\zeta_3 = 6\zeta_{311} + 3\zeta_{221} + \zeta_{212} \,,$$
$$\zeta_2\zeta_{21} = \zeta_{41} + 2\zeta_{32} + 2\zeta_{23} \,. \tag{11.10}$$

The integral representation allows us to derive another set of useful relations, even simpler than shuffling — duality relations. Let's make the substitution $x_i \to 1 - x_i$. Then $\omega_0 \leftrightarrow \omega_1$; in order to preserve the ordering $1 > x_1 > \cdots x_s > 0$, we have to arrange all the ω factors in the opposite order. In other words, after writing down an integral representation for a multiple ζ value, we may read it in the Arabic fashion, from right to left, simultaneously replacing $\omega_0 \leftrightarrow \omega_1$:

$$\zeta_3 = \int \omega_0\omega_0\omega_1 = \int \omega_0\omega_1\omega_1 = \zeta_{21} \,,$$
$$\zeta_4 = \int \omega_0\omega_0\omega_0\omega_1 = \int \omega_0\omega_1\omega_1\omega_1 = \zeta_{211} \,,$$
$$\zeta_5 = \int \omega_0\omega_0\omega_0\omega_0\omega_1 = \int \omega_0\omega_1\omega_1\omega_1\omega_1 = \zeta_{2111} \,,$$
$$\zeta_{41} = \int \omega_0\omega_0\omega_0\omega_1\omega_1 = \int \omega_0\omega_0\omega_1\omega_1\omega_1 = \zeta_{311} \,, \tag{11.11}$$
$$\zeta_{32} = \int \omega_0\omega_0\omega_1\omega_0\omega_1 = \int \omega_0\omega_1\omega_0\omega_1\omega_1 = \zeta_{221} \,,$$
$$\zeta_{23} = \int \omega_0\omega_1\omega_0\omega_0\omega_1 = \int \omega_0\omega_1\omega_1\omega_0\omega_1 = \zeta_{212} \,.$$

Duality relations are the only known relations which say that two distinct multiple ζ values are just equal to each other.

Let's summarize. There are 3 distinct multiple ζ values of weight 4,

$$\zeta_4 = \zeta_{211} \,, \quad \zeta_{31} \,, \quad \zeta_{22} \,,$$

due to duality (11.11). The stuffling (11.5) and shuffling (11.9) relations yield

$$\zeta_2^2 = \zeta_4 + 2\zeta_{22} \quad \to \zeta_{22} = \frac{3}{4}\zeta_4 \,,$$
$$\zeta_2^2 = 4\zeta_{31} + 2\zeta_{22} \to \zeta_{31} = \frac{1}{4}\zeta_4 \,,$$

where we have used

$$\zeta_2^2 = \frac{5}{2}\zeta_4 \,.$$

This follows from

$$\zeta_2 = \frac{\pi^2}{6}, \quad \zeta_4 = \frac{\pi^4}{90}. \tag{11.12}$$

Therefore, all multiple ζ values of weight 4 can be expressed via ζ_4.

There are 4 distinct multiple ζ values of weight 5,

$$\zeta_5 = \zeta_{2111}, \quad \zeta_{41} = \zeta_{311}, \quad \zeta_{32} = \zeta_{221}, \quad \zeta_{23} = \zeta_{212},$$

due to duality (11.11). The stuffling (11.5) and shuffling (11.10) relations yield

$$\begin{cases} \zeta_2 \zeta_3 = \zeta_{32} + \zeta_{23} + \zeta_5, \\ \zeta_2 \zeta_3 = 6\zeta_{41} + 3\zeta_{32} + \zeta_{23}, \\ \zeta_2 \zeta_3 = \zeta_{41} + 2\zeta_{32} + 2\zeta_{23}. \end{cases}$$

Solving this linear system, we obtain

$$\zeta_{41} = 2\zeta_5 - \zeta_2\zeta_3, \quad \zeta_{32} = -\frac{11}{2}\zeta_5 + 3\zeta_2\zeta_3, \quad \zeta_{23} = \frac{9}{2}\zeta_5 - 2\zeta_2\zeta_3. \tag{11.13}$$

Therefore, all multiple ζ values of weight 5 can be expressed via ζ_5 and $\zeta_2\zeta_3$.

Weight 5 is sufficient for calculating finite parts of three-loop propagator diagrams. We shall not discuss higher weights here.

11.2   Expanding hypergeometric functions in ε: an example

There is an algorithm to expand in ε hypergeometric functions of unit argument whose indices tend to integers at $\varepsilon \to 0$. The coefficients are linear combinations of multiple ζ values. We shall first consider an example: expanding the $_3F_2$ function (8.23), which appears in $G(1,1,1,1,2+\varepsilon)$ (Sect. 8.5), up to $\mathcal{O}(\varepsilon)$.

The function we want to expand is

$$F = \sum_{n=0}^{\infty} \frac{(2 - 2\varepsilon)_n (2 + 2\varepsilon)_n}{(3 + \varepsilon)_n (3 + 2\varepsilon)_n}. \tag{11.14}$$

When arguments of two Pochhammer symbols differ by an integer, their ratio can be simplified:

$$\frac{(2 + 2\varepsilon)_n}{(3 + 2\varepsilon)_n} = \frac{2 + 2\varepsilon}{n + 2 + 2\varepsilon}.$$

Step 1. We rewrite all Pochhammer symbols $(m + l\varepsilon)_n$ via $(1 + l\varepsilon)_{n+m-1}$:

$$(2 - 2\varepsilon)_n = \frac{(1 - 2\varepsilon)_{n+1}}{1 - 2\varepsilon}, \quad (3 + \varepsilon)_n = \frac{(1 + \varepsilon)_{n+2}}{(1 + \varepsilon)(2 + \varepsilon)}.$$

Then our function F (11.14) becomes

$$F = \frac{2(2 + \varepsilon)(1 + \varepsilon)^2}{1 - 2\varepsilon} F' = 2(2 + 9\varepsilon + \cdots)F',$$

$$F' = \sum_{n=0}^{\infty} \frac{1}{n + 2 + 2\varepsilon} \frac{(1 - 2\varepsilon)_{n+1}}{(1 + \varepsilon)_{n+2}}. \tag{11.15}$$

In what follows, we shall consider F', and return to F at the end of calculation. It is convenient to introduce the function $P_n(\varepsilon)$:

$$(1 + \varepsilon)_n = n!\, P_n(\varepsilon), \quad P_n(\varepsilon) = \prod_{n'=1}^{n} \left(1 + \frac{\varepsilon}{n'}\right). \tag{11.16}$$

Then

$$F' = \sum_{n=0}^{\infty} \frac{1}{(n + 2)(n + 2 + 2\varepsilon)} \frac{P_{n+1}(-2\varepsilon)}{P_{n+2}(\varepsilon)}.$$

Step 2. We expand the rational function in front of P's in ε:

$$\frac{1}{(n + 2)(n + 2 + 2\varepsilon)} = \frac{1}{(n + 2)^2} - \frac{2\varepsilon}{(n + 2)^3} + \cdots$$

In this simple case, all n-dependent brackets are the same $(n + 2)$; in more general cases, we should have to decompose each term into partial fractions with respect to n. Then

$$F' = \sum_{n=0}^{\infty} \frac{1}{(n + 2)^2} \frac{P_{n+1}(-2\varepsilon)}{P_{n+2}(\varepsilon)} - 2\varepsilon \sum_{n=0}^{\infty} \frac{1}{(n + 2)^3} + \mathcal{O}(\varepsilon^2),$$

because we may replace all $P_n(l\varepsilon) \to 1$ in the $\mathcal{O}(\varepsilon)$ term. Shifting the summation indices, we get

$$F' = \sum_{n=2}^{\infty} \frac{1}{n^2} \frac{P_{n-1}(-2\varepsilon)}{P_n(\varepsilon)} - 2\varepsilon \sum_{n=2}^{\infty} \frac{1}{n^3} + \mathcal{O}(\varepsilon^2).$$

Step 3. We rewrite all $P_{n+m}(l\varepsilon)$ via $P_{n-1}(l\varepsilon)$:

$$P_n(\varepsilon) = \left(1 + \frac{\varepsilon}{n}\right) P_{n-1}(\varepsilon).$$

We again expand in ε (and, if necessary, decompose into partial fractions) rational functions in front of P's:

$$F' = \sum_{n=2}^{\infty} \frac{1}{n^2} \frac{P_{n-1}(-2\varepsilon)}{P_{n-1}(\varepsilon)} - 3\varepsilon \sum_{n=2}^{\infty} \frac{1}{n^3} + \mathcal{O}(\varepsilon^2).$$

Step 4. We add and subtract a few first terms to have all summations start from $n = 1$:

$$F' = \sum_{n=1}^{\infty} \frac{1}{n^2} \frac{P_{n-1}(-2\varepsilon)}{P_{n-1}(\varepsilon)} - 1 - 3(\zeta_3 - 1)\varepsilon + \mathcal{O}(\varepsilon^2).$$

Step 5. We expand $P_{n-1}(l\varepsilon)$ in ε:

$$P_{n-1}(\varepsilon) = \prod_{n>n'>0} \left(1 + \frac{\varepsilon}{n'}\right) = 1 + z_1(n)\varepsilon + z_{11}(n)\varepsilon^2 + \cdots \qquad (11.17)$$

where

$$z_s(n) = \sum_{n>n_1>0} \frac{1}{n_1^s},$$
$$z_{s_1 s_2}(n) = \sum_{n>n_1>n_2>0} \frac{1}{n_1^{s_1} n_2^{s_2}}, \qquad (11.18)$$

and so on. These finite z sums obey the same stuffling relations as infinite ζ sums, for example,

$$z_s(n)z_{s_1 s_2}(n) = z_{s s_1 s_2}(n) + z_{s+s_1, s_2}(n) + z_{s_1 s s_2}(n) + z_{s_1, s+s_2}(n) + z_{s_1 s_2 s}(n). \qquad (11.19)$$

Therefore, coefficients of expansion of ratios of products of P functions can always be made linear in z sums. In our simple case,

$$\frac{P_{n-1}(-2\varepsilon)}{P_{n-1}(\varepsilon)} = \frac{1 - 2z_1(n)\varepsilon + \cdots}{1 + z_1(n)\varepsilon + \cdots} = 1 - 3z_1(n)\varepsilon + \cdots$$

Therefore,

$$F' = \sum_{n>0} \frac{1}{n^2} - 3\varepsilon \sum_{n>n_1>0} \frac{1}{n^2 n_1} - 1 - 3(\zeta_3 - 1)\varepsilon + \cdots$$
$$= \zeta_2 - 1 - 3(\zeta_{21} + \zeta_3 - 1)\varepsilon + \cdots$$

Recalling the duality relation (11.11) $\zeta_{21} = \zeta_3$, we obtain for our function F (11.14)

$$F = 2(2 + 9\varepsilon + \cdots)\left[\zeta_2 - 1 - 3(2\zeta_3 - 1)\varepsilon + \cdots\right]$$
$$= 4(\zeta_2 - 1) + 6(-4\zeta_3 + 3\zeta_2 - 1)\varepsilon + \cdots$$

Thus we have reproduced the first two terms of (8.23).

11.3 Expanding hypergeometric functions in ε: the algorithm

Now we shall formulate the algorithm in a general setting. We want to expand

$$F = \sum_{n=0}^{\infty} \frac{\prod_i (m_i + l_i\varepsilon)_n}{\prod_{i'}(m'_{i'} + l'_{i'}\varepsilon)_n} \tag{11.20}$$

in ε.

Step 1. We rewrite all Pochhammer symbols $(m + l\varepsilon)_n$ via $(1 + l\varepsilon)_{n+m-1}$:

$$(m + l\varepsilon)_n = \frac{(1 + l\varepsilon)_{n+m-1}}{(1 + l\varepsilon)\cdots(m - 1 + l\varepsilon)}.$$

Introducing the function P (11.16), we rewrite F (11.20) as

$$F = \sum_{n=0}^{\infty} R(n, \varepsilon) \frac{\prod_i P_{n+m_i-1}(l_i\varepsilon)}{\prod_{i'} P_{n+m'_{i'}-1}(l'_{i'}\varepsilon)},$$

where R is a rational function of ε and n.

Step 2. We expand

$$R(n, \varepsilon) = R_0(n) + R_1(n)\varepsilon + \cdots$$

and decompose each $R_j(n)$ into partial fractions with respect to n. There is a catch here: partial-fractioning of a convergent series can split it into a combination of logarithmically divergent ones. Therefore, it is necessary to keep the upper limit of summation finite, and go to the limit only after combining such series together. Because of shifts of the summation variable, cancellation of divergent series leaves us with a few terms around the upper limit. They, however, tend to zero, and this means that we may formally manipulate logarithmically divergent series as if they were convergent.

Shifting summation indices, we can write F (11.20) as a sum of terms of the form

$$\sum_{n=n_0}^{\infty} \frac{1}{n^k} \frac{\prod_i P_{n+m_i}(l_i\varepsilon)}{\prod_{i'} P_{n+m'_{i'}}(l'_{i'}\varepsilon)} .$$

Step 3. We rewrite all $P_{n+m}(l\varepsilon)$ via $P_{n-1}(l\varepsilon)$:

$$P_{n+m}(l\varepsilon) = P_{n-1}(l\varepsilon) \times \left(1 + \frac{l\varepsilon}{n}\right) \cdots \left(1 + \frac{l\varepsilon}{n+m}\right) .$$

Our F becomes a sum of terms of the form

$$\sum_{n=n_0}^{\infty} \frac{R(n,\varepsilon)}{n^k} \frac{\prod_i P_{n-1}(l_i\varepsilon)}{\prod_{i'} P_{n-1}(l'_{i'}\varepsilon)} ,$$

where R is a rational function of ε and n. Moving from terms with lower powers of ε to ones with higher powers, we expand R's in ε:

$$R(n,\varepsilon) = 1 + R_1(n)\varepsilon + \cdots$$

Terms of higher orders in ε are decomposed in partial fractions again. Finally, F becomes a sum of terms of the form

$$\sum_{n=n_0}^{\infty} \frac{1}{n^k} \frac{\prod_i P_{n-1}(l_i\varepsilon)}{\prod_{i'} P_{n-1}(l'_{i'}\varepsilon)} .$$

Step 4. Adding and subtracting terms with n from 1 to $n_0 - 1$, we rewrite F as a sum of terms of the form

$$\sum_{n=1}^{\infty} \frac{1}{n^k} \frac{\prod_i P_{n-1}(l_i\varepsilon)}{\prod_{i'} P_{n-1}(l'_{i'}\varepsilon)}$$

and rational functions of ε (which can be trivially expanded in ε).

Step 5. We expand each $P_{n-1}(l\varepsilon)$ as (11.17). Expansions of ratios of products of P functions contain products of z sums; they are reduced to linear combinations of z sums by stuffling relations (e.g., (11.19)). Using

$$\sum_{n=1}^{\infty} \frac{1}{n^k} z_{s_1\ldots s_j}(n) = \zeta_{ks_1\ldots s_j} ,$$

we can express coefficients of ε expansion of F as linear combinations of multiple ζ values (discussed in Sect. 11.1) and rational numbers. *Q. E. D.*

A few historical comments. In 2000, I needed to expand some $_3F_2$ functions of unit argument in ε (in fact, they were (9.32) and (9.40)). I asked

David Broadhurst how to do this, and he replied: just expand Pochhammer symbols in ε, the coefficients will be expressible via multiple ζ values. Following this advise, I implemented the algorithm described above in RE-DUCE, and obtained the expansions (9.32) and (9.40). Presentation in Sects. 11.2 and 11.3 closely follows my notes from 2000. Later this algorithm was described as Algorithm A in [Moch *et al.* (2002)] (this paper also contains more difficult algorithms (B, C, D) to expand some more complicated sums in ε). All of these algorithms are implemented in the C++ library nestedsums [Weinzierl (2002)][1]. It is based on the C++ computer algebra library GiNaC [Bauer *et al.* (2002)].

The problem of expanding in ε hypergeometric functions some of whose indices tend to half-integers at $\varepsilon \to 0$ is more difficult. Such functions appear in on-shell propagator calculations, see Sects. 10.2, 10.3. This problem is discussed in recent papers [Weinzierl (2004); Davydychev and Kalmykov (2004); Kalmykov (2006)]. Such expansions contain many constants more complicated than multiple ζ values.

---

[1]Unfortunately, there exist different notations for multiple ζ values. We follow [Borwein *et al.* (2001)] here; in [Moch *et al.* (2002); Weinzierl (2002)] the order of indices is reversed.

Bibliography

M. Argeri, P. Mastrolia, E. Remiddi, Nucl. Phys. B **631** (2002) 388

D.T. Barfoot, D.J. Broadhurst, Z. Phys. C **41** (1988) 81

C. Bauer, A. Frink, R. Kreckel, J. Symbolic Computations **33** (2002) 1; cs.SC/0004015; http://www.ginac.de/

M. Beneke, V.M. Braun, Nucl. Phys. B **426** (1994) 301

I. Bierenbaum, S. Weinzierl, Eur. Phys. J. C **32** (2003) 67

J.M. Borwein, D.M. Bradley, D.J. Broadhurst, P. Lisoněk, Trans. Amer. Math. Soc. **355** (2001) 907, math.CA/9910045

D.J. Broadhurst, Z. Phys. C **32** (1986) 249

D.J. Broadhurst, Z. Phys. C **47** (1990) 115

D.J. Broadhurst, Z. Phys. C **54** (1992) 599

D.J. Broadhurst, hep-th/9604128

D.J. Broadhurst, in *New computing techniques in physics research II*, ed. D. Perret-Gallix, World Scientific (1992), p. 579

D.J. Broadhurst, J.A. Gracey, D. Kreimer, Z. Phys. C **75** (1997) 559

D.J. Broadhurst, N. Gray, K. Schilcher, Z. Phys. C **52** (1991) 111

D.J. Broadhurst, A.G. Grozin, Phys. Lett. B **267** (1001) 105

D.J. Broadhurst, A.G. Grozin, Phys. Rev. D **52** (1995) 4082

D.J. Broadhurst, A.G. Grozin, in *New Computing Techniques in Physics Research IV*, ed. B. Denby, D. Perret-Gallix, World Scientific (1995), p. 217

K.G. Chetyrkin, A.G. Grozin, Nucl. Phys. B **666** (2003) 289; http://www-ttp.physik.uni-karlsruhe.de/Progdata/ttp03/ttp03-10/

K.G. Chetyrkin, A.L. Kataev, F.V. Tkachev, Nucl. Phys. B **174** (1980) 345

K.G. Chetyrkin, F.V. Tkachov, Nucl. Phys. B **192** (1981) 159; F.V. Tkachov, Phys. Lett. B **100** (1981) 65

A. Czarnecki, A.G. Grozin, Phys. Lett. B **405** (1997) 142

A. Czarnecki, K. Melnikov, Phys. Rev. D **66** (2002) 011502

A.I. Davydychev, A.G. Grozin, Phys. Rev. D **59** (1999) 054023

A.I. Davydychev, M.Yu. Kalmykov, Nucl. Phys. B **699** (2004) 3

J. Fleischer, M.Yu. Kalmykov, Comput. Phys. Commun. **128** (2000) 531

J. Fleischer, O.V. Tarasov, Phys. Lett. B **283** (1992) 129; Comput. Phys. Commun. **71** (1992) 193

S.G. Gorishnii, A.P. Isaev, Theor. Math. Phys. **62** (1985) 232

S.G. Gorishny, S.A. Larin, F.V. Tkachov, Preprint INR P-0330, Moscow (1984); S.G. Gorishny, S.A. Larin, L.R. Surguladze, F.V. Tkachov, Comput. Phys. Commun. **55** (1989) 381

N. Gray, D.J. Broadhurst, W. Grafe, K. Schilcher, Z. Phys. C **48** (1990) 673

A.G. Grozin, Using REDUCE in High Energy Physics, Cambridge University Press (1997)

A.G. Grozin, J. High Energy Phys. **03** (2000) 013, hep-ph/0002266; http://www-ttp.physik.uni-karlsruhe.de/Progdata/ttp00/ttp00-01/

A.G. Grozin, Heavy Quark Effective Theory, Springer tracts in modern physics **201**, Springer (2004)

A.C. Hearn, REDUCE User's Manual, Version 3.8 (2004)

M.Yu. Kalmykov, J. High Energy Phys. **04** (2006) 056

D.I. Kazakov, Theor. Math. Phys. **58** (1984) 223

D.I. Kazakov, Theor. Math. Phys. **62** (1985) 84

A.V. Kotikov, Phys. Lett. B **375** (1996) 240

S. Laporta, E. Remiddi, Phys. Lett. B **379** (1996) 283

S.A. Larin, F.V. Tkachov, J.A.M. Vermaseren, Preprint NIKHEF-H/91-18, Amsterdam (1991)

A.V. Manohar, M.B. Wise, *Heavy Quark Physics*, Cambridge University Press (2000)

K. Melnikov, T. van Ritbergen, Phys. Rev. Lett. **84** (2000) 1673

K. Melnikov, T. van Ritbergen, Phys. Lett. B **482** (2000) 99

K. Melnikov, T. van Ritbergen, Nucl. Phys. B **591** (2000) 515

S. Moch, P. Uwer, S. Weinzierl, J. Math. Phys. **43** (2002) 3363

M. Neubert, Phys. Reports **245** (1994) 259

A.I. Onishchenko, O.L. Veretin, Phys. Atom. Nucl. **68** (2005) 1405

V.A. Smirnov, *Feynman Integral Calculus*, Springer (2006)

M. Veltman, SCHOONSCHIP, CERN (1967); H. Strubbe, Comput. Phys. Commun. **8** (1974) 1

J.A.M. Vermaseren, Symbolic Manipulation with FORM, Amsterdam (1991); math-ph/0010025; http://www.nikhef.nl/~form

S. Weinzierl, Comput. Phys. Commun. **145** (2002) 357; http://wwwthep.physik.uni-mainz.de/~stefanw/nestedsums/

S. Weinzierl, J. Math. Phys. **45** (2004) 2656

Index